*Electronic
Analog-to-Digital
Converters*

Electronic Analog-to-Digital Converters

Principles, Circuits, Devices, Testing

Dr-Ing. Dieter Seitzer
Professor of Electronic Engineering,
University of Erlangen-Nuremberg, West Germany.

Dr-Ing. Günter Pretzl
University of Erlangen-Nuremberg, West Germany.

Dr-Ing. Nadder A. Hamdy
University Assiut, Egypt.

A Wiley–Interscience Publication

JOHN WILEY & SONS
Chichester · New York · Brisbane · Toronto · Singapore

Copyright © 1983, by John Wiley & Sons Ltd.

All rights reserved.

No part of this book may be reproduced by any means, nor transmitted, nor translated into a machine language without the written permission of the publisher.

Library of Congress Cataloging in Publication Data:

Seitzer, Dieter, 1933–
   Electronic analog-to-digital converters.

   'A Wiley–Interscience publication.'
   Includes index.
    1. Analog-to-digital converters. 2. Digital electronics. I. Pretzl, Günter. II. Hamdy, Nadder.
III. Title.
TK7887.6.S437   1984   621.3819'596   83-7037
ISBN 0 471 90198 9

British Library Cataloguing in Publication Data:
Seitzer, Dieter
   Electronic analog-to-digital converters.
   1. Digital-to-analog converters 2. Analog-to-digital converters
   I. Title II. Pretzl, Günter
   III. Hamdy, Nadder, A.
   621.3819'596   TK7887.6

   ISBN 0 471 90198 9

Typeset by Pintail Studios Ltd, Ringwood, Hampshire.
Printed by the Pitman Press, Bath, Avon.

Contents

Preface		ix
1. Introduction		1
1.1	What are Analog-to-Digital Converters?	2
1.2	Applications of Analog-to-Digital Converters	3
1.3	Performance Characterization	4
1.4	Definition of the Conversion Problem	6
	References	7
2. Underlying Principles		8
2.1	The Concept of Pulse Code Modulation	8
2.2	Codes for Analog-to-Digital Conversion	11
2.3	Efficient Coding	15
2.3.1	Coding of Independent Messages	16
2.3.2	Predictive Coding	18
	References	22
3. Analog-to-Digital Conversion Techniques		24
3.1	Classification of the Methods	24
3.2	The Parallel (Direct) Converter Structure	24
3.2.1	The Cathode Ray Encoding Tube	26
3.2.2	Parallel Converters with Voltage Comparison	28
3.2.3	A Parallel Converter with Current Comparison	33
3.3	Cascaded Converters	34
3.3.1	Introduction	34
3.3.2	Sequential-Parallel Converters as an Extension of Direct Converters	35
3.3.3	Cascaded 1-Bit per Stage Converters	40
3.3.4	Gray Code Converters	41
3.3.5	Pipelining Analog-to-Digital Converters	49
3.4	Feedback Analog-to-Digital Converters	50
3.4.1	Introduction	50
3.4.2	The Counting Method and its Extension	53
3.4.3	Circulation 1-Bit per Cycle Analog-to-Digital Converters	54
3.4.4	Examples of Multichip or Monolithic Feedback Converters	56
3.4.5	Interpolation Analog-to-Digital Converters	58

3.5	Indirect Converters		61
	3.5.1 The Single-Ramp Method		61
	3.5.2 Dual-Ramp Converters		63
	3.5.3 Voltage-to-Frequency Converters		65
	3.5.4 Charge Redistribution Converters		67
3.6	Miscellaneous Analog-to-Digital Converters		69
	3.6.1 The Put and Take Technique		69
	3.6.2 Non-linear Analog-to-Digital Converters		70
	3.6.3 Floating Point Analog-to-Digital Converters		75
	3.6.4 Multiplexed Analog-to-Digital Converters		83
	3.6.5 Delta Modulators		84
	3.6.6 Stochastic Analog-to-Digital Converters		86
	3.6.7 Ultra-High-Speed Analog-to-Digital Conversion		89
	3.6.8 Transient Analog-to-Digital Conversion		90
3.7	Analog-to-Digital Conversion and Microcomputers		92
	3.7.1 Introduction		92
	3.7.2 Analog-to-Digital Conversion by Software		92
	3.7.3 Data Acquisition Building Blocks		96
	3.7.4 Vertical Analog-to-Digital Conversion		97
3.8	Summary and Conclusions		97
	References		102

4. Digital-to-Analog Converters ... 107

4.1	Introduction	107
4.2	Weighted Current-Digital-to-Analog Converters	108
4.3	Ladder Network Digital-to-Analog Converters	110
4.4	Voltage Divider Digital-to-Analog Converters	114
4.5	Non-linear Digital-to-Analog Converters	114
4.6	Multiplying Digital-to-Analog Converters	117
4.7	Error Correction in Digital-to-Analog Converters	119
4.8	The Shannon–Rack Decoder	120
4.9	Charge Redistribution Digital-to-Analog Converters	121
4.10	A Note on Stochastic and Interpolative Digital-to-Analog Converters	122
	References	122

5. Devices and Building Blocks for Analog-to-Digital Converters ... 124

5.1	Switches		124
	5.1.1 Mechanical Switches		125
	5.1.2 Semiconductor Diodes		126
	5.1.3 The Bipolar Transistor Inverted Mode Switch		127
	5.1.4 The Saturated Emitter Follower		131
	5.1.5 The Current Switch Circuit		131
	5.1.6 Opto-electronic Switches		133

		5.1.7	The Field Effect Transistor as a Switch	133
		5.1.8	Analog Switches Using Diodes	138
	5.2	Sample-and-Hold Circuits		139
		5.2.1	Introduction	139
		5.2.2	A Two-Phase Switch	142
		5.2.3	The Diode Bridge Sample-and-Hold Circuit	144
		5.2.4	Monolithic Sample-and-Hold Circuits	147
		5.2.5	Ultra-High-Speed Sample-and-Hold Circuitry	148
		5.2.6	A Digital Sample-and-Hold Circuit	149
	5.3	Comparators		151
		5.3.1	Introduction	151
		5.3.2	Bandwidth and Transient Characteristics	152
		5.3.3	Operational Amplifiers as Comparators	156
		5.3.4	Tunnel Diode Comparators	159
		5.3.5	The Josephson Junction as a Comparator	161
	5.4	Voltage Reference Circuits		162
		5.4.1	A Zener Diode Reference Circuit	162
		5.4.2	The Bandgap Reference	163
	5.5	Sawtooth Generators		164
		5.5.1	Introduction	164
		5.5.2	The Miller Integrator	165
		5.5.3	The Bootstrap Integrator	166
		5.5.4	Switched Constant Current Charging	167
		5.5.5	Staircase Generators	170
	5.6	Counters		171
		5.6.1	Asynchronous (Ripple) Counters	171
		5.6.2	Synchronous Counters	173
		5.6.3	Up/Down Counters	173
		References		174
6.	Testing Converters			177
	6.1	Introduction		177
	6.2	Measurements on Digital-to-Analog Converters		177
		6.2.1	Static Testing of Digital-to-Analog Converters	180
		6.2.2	Dynamic Testing of Digital-to-Analog Converters	181
	6.3	Measurements on Analog-to-Digital Converters		186
		6.3.1	Static Testing of Analog-to-Digital Converters	189
		6.3.2	Dynamic Testing of Analog-to-Digital Converters	197
	6.4	Summary		213
		References		215

Appendix: Glossary of Terms ... 217

Index ... 221

Preface

The field of A/D and D/A conversion is rapidly developing. It has a long history and will continue to be of interest due to the interdisciplinary links between instrumentation, communications, and computers from an applications point of view, and between circuit devices and semiconductor processing from the design and technology viewpoint.

At any moment in time it is difficult to assess the state of the art, and the danger is that the wheel can be re-invented. Although many articles can be found in the literature, there are few that are comprehensive. In this book an attempt is made to develop a both comprehensive and comprehensible overview of the state of the art at the beginning of the 1980s.

Originally conceived as a third-year textbook for master degree students, its coverage of the subject is equivalent to a three-month, 2 hours per week course. With its many references to contemporary literature, especially IEEE conferences and journals, it should also serve as a quick reference for the practising engineer, as well as for newcomers from related disciplines. It would also be a contribution to the promotion of the field as a whole if the terminology and classification of A/D converters used in this book created a mutual understanding between designers and users of converters.

The authors have tried to put the text on paper while being very much aware of the movement of the target they were aiming at. So the method of subject selection should be debated, and constructive criticism from the reader is welcome.

At the same time, the authors gratefully acknowledge the dedication that Mrs Gerhäuser gave to the text processing and drawings. Last, but not least, we thank the publishers for their patience in view of the various delays due to other commitments on the part of the authors.

Erlangen, December 1982   D. SEITZER
G. PRETZL
N. A. HAMDY

CHAPTER 1

Introduction

Analog-to-digital converters are bridging the gap between the traditional analog signal-sources and digital devices or systems. Due to the still-increasing application of digital signal processing, transmission, storage, control, and display, data converters are also of growing significance. On the systems level, the trend now is towards 'all-digital' realization. This is supported by the widespread use of data-processing equipment that provides flexible and, due to increasing standardization, economic means of handling information.

In the field of signal processing digital methods are also in progress. This is due to the fact that the recently developed technologies for the manufacture of digital integrated circuits (e.g. monolithic LSI circuits) have resulted in a drastic reduction in their costs, a good example being the reduction in the price of electronic calculators over the last decade. In addition, the proliferation of microprocessors and the successful introduction of international standards for interfacing circuits such as the IEC–BUS (IEEE–BUS) and CAMAC standard (Computer-Aided Measurements and Control) are further indications of the trend towards digital methods.[1–3]

Sources of information such as microphones and television cameras and transducers of physical quantities such as strain, temperature, and pressure, etc. deliver essentially continuous quantities, i.e. analog signals. It is therefore the purpose of analog-to-digital conversion to provide the necessary link to digital systems wherever these signals are to be processed, stored, and/or transmitted on a digital medium. The main design objective is to find a reasonable trade-off between the accuracy required and the speed of the system while keeping the costs acceptable. In this way the use of relatively cheap digital circuitry for processing can compensate for the additional cost of the A/D converter. Therefore digital solutions can compete with conventional ones in terms of cost unless use of the latter is unavoidable.

After introducing the subject in Chapter 1 some basic principles of signal theory and definitions are given in Chapter 2. The different methods and techniques employed in data converters are discussed in Chapter 3, their classification being based on parameters of architecture or hardware structure. Examples representing the current state of the art are given throughout the text. In Chapter 4 the reverse process, i.e. digital-to-analog (D/A) conversion, is considered. The various building blocks and components of A/D converters are dealt with in Chapter 5. Finally, the presentation of the subject is completed in Chapter 6, which describes the testing and measurement of A/D converters.

1.1 What are Analog-to-Digital Converters?

A/D converters link the analog (physical) world to the domain of discrete numbers and computers. From a hardware point of view they belong to the category of interface equipment. A/D converters can also be considered as belonging to the family of instrumentation devices (Fig. 1.1). For example, while the pointer reading of a voltmeter in an analog representation is continuous, the reading in a digital representation is discrete, i.e. it is limited to a finite set of values (numbers). The circuitry in digital systems generally uses binary numbers and the reading can be displayed in the familiar decimal form.

In Fig. 1.2(a) continuous and discrete forms of representation are compared. It is obvious that the reading on the left scale covers all possible values within the

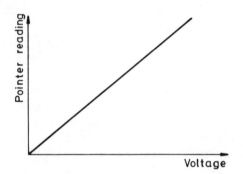

Fig. 1.1 Analog representation: pointer reading versus the measured quantity (voltage) is continuous

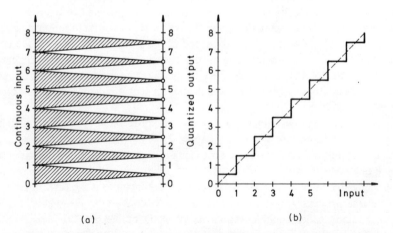

(a) (b)

Fig. 1.2 Quantized representation: allocation of intervals to a number of discrete levels. (a) Allocation of input to output scale (example); (b) input/output relationship as a transfer characteristic

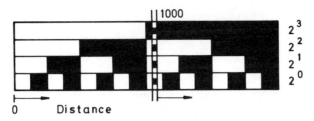

Fig. 1.3   Digital position encoder

whole range, whereas only eight discrete levels can be assumed on the output scale. Therefore it must be concluded that such a discrete reading is associated with an error whose worst-case value may amount to half the size of a step of quantization. These inferences can also be taken from the input/output characteristic of the quantizer illustrated in Fig. 1.2(b).

As a further illustration the process of mechanical position encoding into binary form is considered. As shown in Fig. 1.3, a movable ruler slides below four brushes that scan its segments, connected to a voltage when dark or to ground when light. The four segments are subdivided according to the coefficients of the binary number system. Thus detection of a voltage represents a logical '1', no voltage a binary '0'. All four digits are encoded at once, i.e. in parallel (possible errors are treated later). More modern versions make use of opto-electronic devices instead of mechanic segment scanning.

1.2   Applications of Analog-to-Digital Converters

The most popular application of A/D converters is in the field of digital multimeters (DMM), where the magnitude of a voltage, current, or resistance is directly displayed in decimal form. In addition, they are used extensively in applications such as data acquisition for industrial instrumentation, telemetry, aerospace systems, hybrid computation, traffic control, communications, and numerical control of industrial processes.

In a telemetry system, for example, signals from different measuring stations (transducers) are multiplexed before being converted (Fig. 1.4). Centralized processing takes place at the receiving end. Data converters are used in deep space missions where physical data can only be recorded or transmitted digitally.

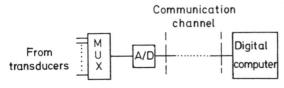

Fig. 1.4   Block diagram of a typical telemetry system

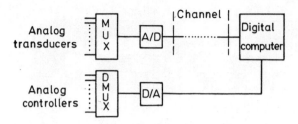

Fig. 1.5 Elements of a digital control system

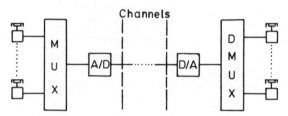

Fig. 1.6 Block diagram of a digital telephony system based on pulse code modulation (PCM)

On-line control of multi-variable processes includes, beside data converters, a digital controller (a digital computer or microprocessor system) that receives the measured data and processes them according to a stored program. The controller responds by instructions in the form of controlling (correction) signals. These instructions are then converted back to analog before they are demultiplexed (Fig. 1.5) to feed the individual actuators. Applications of this scheme are to be found in the chemical industry (high furnaces), steel production, motor control in transportation systems, and communications as well as in consumer electronics and medicine. A vast field of application can be found in digital telephone systems under the name of pulse code modulation (PCM) (Fig. 1.6). In fact, the list of current applications of data converters is increasing rapidly, and the reader is referred to the selected references given at the end of this chapter.

1.3 Performance Characterization

Familiarity with the specifications and performance characteristics of A/D converters is of fundamental importance in selecting a suitable converter for a specific application and/or for comparing different realizations.

The main specifications of an A/D converter can be grouped into three sets: the analog input, the digital output, and the relationship between the two. The first group of data deals with the form of the analog input signal (voltage, current, etc.), and its range (minimum, maximum, bipolar, unipolar). Another group of specifications concerns the digital output, and indicates the output word length (i.e. the number of bits (digits) in each word), the electrical output voltage level, the code used (straight binary, BCD, Gray, etc.), the format of the digital output (i.e. serial on one line or parallep over n-lines) and the resulting bit rate. The length of the digital word represents, in fact, the number of voltage levels the converter can discriminate and hence it is considered to be direct measure of converter resolution as well as accuracy, which should correspond but rarely does. A word length of 12 bits (i.e. 4096 levels) is now considered a standard value. Converters having a resolution of 14 to 16 bits are available in a limited variety, and even 20 bits are considered feasible.

Various sources of errors are associated with the conversion process. These are illustrated in Fig. 1.7, where the transfer characteristic (input/output relationship) of an ideal converter is compared with that of a real one. The ideal quantization levels are shown to be equidistant over the whole range, and hence the average slope is unity. The real characteristic, however, does not start at an input level corresponding to one quantization interval as in the case of the ideal level; hence the converter is said to have an offset error. If adjacent intervals are not equal in size, the average slope becomes non-linear. The related error is the differential non-linearity, which may lead to a missing (skipped) code and a monotonicity error. These errors also result in a gain error (see Chapter 6).

Another important parameter of an A/D converter is the conversion speed. This is given by the number of converted output words per second (conversion rate) or, alternatively, by the time needed for a single conversion (conversion time). However, usually the two values are not exactly reciprocal. The number of

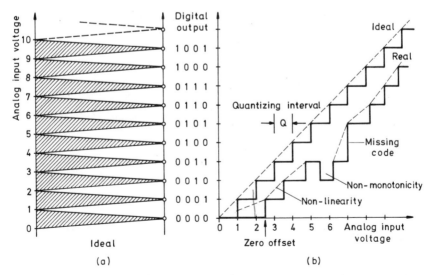

Fig. 1.7 Input/output relationship of an A/D converter. (a) Digital representation of analog input range; (b) transfer characteristic of an ideal and a real converter

conversions per second can be either smaller or larger than the reciprocal of the time for a single conversion, depending on the converter structure. Values of more than 10^7 conversions/second at a word length of 10 bits represent two limits that are feasible but difficult to exceed within reasonable cost.[4] 'Reasonable' here means that the price of the set-up designed for a certain application has to be acceptable from the user's point of view. Speed and cost parameters are very sensitive to technological progress. Gigabit per second output rates[5] are currently at the research stage, and cost becomes less stringent the more data converters appear as subsystems on VLSI chips for data acquisition, signal processors, or communications system building blocks.[6] More details of A/D converter performance characterization and errors as well as their exact definition are given in Chapter 6.

1.4 Definition of the Conversion Problem

For conversion of an analog signal a system with the basic building blocks is used Fig. 1.8). First, the frequency band of the signal must be limited by an antialiasing, i.e. low-pass, filter. The sample-and-hold circuit (S/H) must take samples periodically from the analog input signal and maintain the instantaneously obtained amplitude constant for a certain period. During this hold period the A/D converter performs its main functions, namely quantizing and encoding. The proper sequence of functions is achieved by means of a control logic. After a certain delay time, the digital output signal is available.

To summarize: the function of an A/D converter is to create a discrete signal both in time and amplitude from the originally continuous signal and then to assign the obtained discrete amplitudes to a desired code. In the following, only the binary code will be considered unless otherwise mentioned.

If the output signal s_a of a quantizer, i.e. an A/D converter linked with a D/A converter, is normalized such that the quantizing interval size is unity ($q = 1$), then:

$$s_a = b_{n-1} \cdot 2^{n-1} + b_{n-2} \cdot 2^{n-2} + \ldots + b_k 2^k + \ldots + b_0 2^0$$

where $b_k = 0$ or 1 is the binary weight or coefficient, n is the length of the digital word, b_{n-1} is the *M*ost *S*ignificant *B*it (MSB), and b_0 is the *L*east *S*ignificant *B*it (LSB).

The major task is then to determine the values of the n bits of one code word.

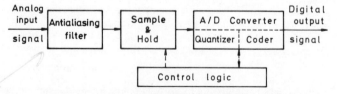

Fig. 1.8 Fundamental structure for A/D conversion

The quantized signal covers the range

$$0 \leqslant s_a \leqslant 2^n - 1$$

It can assume $m = 2^n$ values. The full-scale range (FSR) of the input signal s_i is $m \cdot q = 2^n \cdot q$, its maximum power being slightly larger than that of the quantized signal s_a. Frequently, s_a is normalized such that the weight of the MSB becomes 2^{-1}. This is achieved if the above expression is divided by 2^n. The coefficients b_k remain unchanged. In this case, the maximum value of the quantized signal approaches unity:

$$0 \leqslant s'_a = \leqslant 1 - 2^{-n}$$

The building blocks of Fig. 1.8 are explained in the following chapters: various A/D conversion techniques are discussed in Chapter 3 and sample-and-hold circuits in Chapter 5. The structure of control logic circuits are treated in conjunction with section 3.7 on A/D conversion by microcomputers.

References

1. Special issue on microprocessor technology and applications. *Proc. IEEE* **64**, June (1976).
2. Knoblock, D. E., Loughry, D. C., and Vissers, Ch. A., 'Insight into Interfacing', *IEEE Spectrum* **12**, 50–57, May (1975).
3. Horelick, D., and Larsen, R. S., 'CAMAC: A Modular Standard', *IEEE Spectrum* **13**, 50–55, April (1976).
4. Allan, R., 'The Inside News on Data Converters', *Electronics* **53**, 101–12, July 17 (1980).
5. Bosch, B. G., 'Gigabit Electronics—A Review', *Proc. IEEE* **67**, 340–79, March (1979).
6. Mokhoff, N., 'Monolithic Approach bears Fruit in Data Conversion', *Electronics* **52**, 105–16, May (1979).
7. Hoeschele, D. F., *Analog to Digital/Digital to Analog Conversion Techniques*, Wiley, New York (1968).
8. Schmid, H., *Electronic analog/digital conversions*, Van Nostrand Reinhold, New York (1970).
9. Hnatek, E. R., *A User's Handbook of D/A and A/D Converters*, Wiley Interscience, New York (1976).
10. Seitzer, D., *Elektronische Analog-Digital-Umsetzer*, Springer, Berlin/Heidelberg (1977).
11. Sheingold, D. H. (ed.), *Analog-digital Conversion Notes*, Analog Devices Inc., Norwood, Mass. (1977).
12. Special issue on analog-to-digital and digital-to-analog converters. *IEEE Transactions* **CAS-25**, 389–562, July (1978).
13. Tietze, U., and Schenk, Ch., *Advanced Electronic Circuits*, Springer, Berlin/Heidelberg (1978), Chapters 7 and 14.
14. Gordon, B. M. (ed.), *The Analogic Data-Conversion Systems Digest*, 3rd edn, Analogic Corp., Wakefield, Mass. (1978).
15. *Data Acquisition Handbook*, Intersil Inc. (1980).
16. Best, R. E., Eine Systemtheorie der DA- und AD-Converter und ihre Anwendung auf die Konstruktion schneller AD-Converter. PhD thesis, No. 4784, Eidgen. Techn. Hochschule, Zurich (1971).

CHAPTER 2

Underlying Principles

The purpose of this chapter is to discuss the effects of quantization both in time and amplitude on the reconstructed signal. The time interval between the successive samples is related to the analog bandwidth and the number of quantization intervals determines the fidelity of reproduction of the original signal. Encoders that carry out data compression by signal processing are also considered.

2.1 The Concept of Pulse Code Modulation

Digital representation of a signal can be considered as replacing a continuous voltage $V(t)$ (Fig. 2.1(a)) by a periodical sequence of samples (time quantization) whose amplitudes can assume a limited number of levels (amplitude quantization). Detailed analysis of these fundamental relationships is the subject of both signal and system theory.[1] Here we shall limit ourselves to the presentation of some basic results which we need for designing and understanding A/D converters.

Signals occurring within A/D converters are time dependent quantities. An error-free reconstruction of a signal quantized in time is possible if the time interval T_s between the samples is

$$T_s < 1/2B$$

where B is the bandwidth of the analog signal (Fig. 2.1(b)). This relationship is sometimes referred to as the 'sampling theorem'.[2] It means that the sampling or Nyquist freqency $1/T_s = f_s$ should equal at least twice the highest frequency given by the bandwidth B of the signal. In order to give a qualitative explanation we

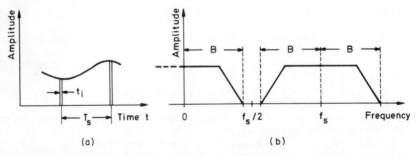

Fig. 2.1 Taking samples from a continuous waveform. (a) Time domain representation; (b) frequency domain representation: spectrum of amplitude modulated pulse sequence

note that the spectrum covered by an amplitude modulated pulse sequence, i.e. the sequence of samples taken from the signal waveform, extends by $\pm B$ from integer multiples of the pulse frequency (Fig. 2.1(b)). Hence to avoid overlapping (aliasing) of adjacent frequency intervals the relationship $f_s > 2B$ can be derived from Fig. 2.1(b). For the same argument, a low-pass filter is connected in front of the sample-and-hold circuit in Fig. 1.8.

The width t_i of the sampling pulses is assumed to be zero in the above discussion. However, the actual case is $t_i > 0$. This will be considered in conjunction with the presentation of the sample-and-hold circuits in section 5.2.

Quantization of the full-range S_i of the input signal into m levels each of a size Q such that

$$S_i = m \cdot Q$$

leads to an information loss, since after D/A conversion the reconstructed signal is associated with an error up to $\pm Q/2$ (Fig. 1.2(b)). Therefore, to keep this unavoidable signal distortion within an acceptable range, m should be chosen as large as possible. However, for economical reasons, m should be as small as possible. One criterion is to select m such that the resulting quantization noise equals the noise produced by other noise sources. However, the decision whether this criterion is usable or not depends on the actual application.

Another approach that is widely accepted is to consider the signal-to-noise ratio (S/N). If all the signal values within the quantizing interval Q are equally probable and the maximum error is $\pm Q/2$, then the dissipated noise power in a resistor R is[2,3]

$$P_n = Q^2/12R$$

The signal-to-noise ratio is defined as follows:

$$(S/N)_{dB} = 10 \cdot \log_{10} P_s/P_n$$

(dB = decibel), where P_s is the signal power. The resulting value depends on the value of the signal power P_s taken as a reference.

If a bipolar analog signal has a triangular waveform and a swing of $\pm m \cdot Q/2$ ($m \gg 1$) the mean signal power is[2]

$$P_m = (mQ)^2/12R$$

Therefore the mean power output signal-to-noise ratio is

$$(S_m/N)_{dB} = 10 \cdot \log_{10} m^2$$

Recalling that in a binary representation with a word length of n bit, i.e. $m = 2^n$, the above expression reduces to

$$(S_m/N)_{dB} = 6 \cdot n$$

Under the same conditions a sine wave has a mean signal power

$$P_m = (mQ)^2/8R$$

This leads to

$$(S_m/N)_{dB} = 10 \log_{10}(1.5m^2) = 6 \cdot n + 1.8$$

A unipolar analog signal such as a video signal has a peak-to-zero signal excursion $S_p = m \cdot Q$. If it is related to the r.m.s. quantizing noise power P_n, the signal-to-noise ratio will be[2]

$$(S_p/N)_{dB} = 20 \log_{10}(2 \cdot \sqrt{3} \cdot m) = 6 \cdot n + 10.8$$

The mean signal power to quantizing noise power ratio S_m/N is given by the relationships mentioned so far. Only for the cases of deterministic type signals such as sine or triangular waves. For real signals, such as voice, sound, etc., the signal range is virtually unlimited, and is defined by $p(x)$, the probability density function (PDF), where $p(x)$ describes the probability of an amplitude of value x.[4] The quantizer performance is usually described in terms of the mean square distortion measure D, where

$$D = \int_{-\infty}^{+\infty} [C(x) - x]^2 \, p(x) dx = Q^2/12$$

$C(x)$ is the quantizer characteristic for which a uniform step size Q and a large number of steps is assumed. The signal-to-noise ratio S_m/N then assumes

$$S_m/N = 10 \log_{10}(\sigma^2/D)$$

where σ^2 is the variance of the input samples, i.e. it is that amplitude leading to the average signal power or r.m.s. value.

Due to the fact that there is a finite range of the quantizer characteristic, there must be a certain level, V, where the signal falls beyond the characteristic, giving rise to overload distortion and noise. To avoid significant overload distortion the overload level is chosen to be a suitable multiple $y = V/\sigma$, called the 'loading factor'. A common choice is $y = 4$. Then the approximate step size is $Q = 8\sigma/m$. For an n-bit quantizer ($m = 2^n$) the signal-to-noise ratio is found to be

$$(S_m/N)_{dB} = 6n - 7.3$$

So the increase of the signal-to-noise ratio for all types of the signals is 6 dB per bit. (The rate is exactly $20 \log_{10} 2 = 6.02$ dB per bit.) The signals differ only by the additive form, which is controlled by the loading factor.

In Fig. 2.2 the dependence of S_m/N is shown for $n = 7$, $m = 128$. The curve shows a maximum S_m/N of about 30 dB. Beyond the corresponding relative input level of -15.7 dB (this means a loading factor of 6.1 or a full-scale range of $12.2 \cdot \sigma$) overload noise causes a deterioration in performance. The curve is based on the assumption of a Laplacian PDF

$$p(x) = \frac{1}{\sqrt{2} \cdot \sigma} \cdot e^{-\{(\sqrt{2} \cdot |x|)/\sigma\}}$$

(This has been normalized so that the mean is zero and the variance σ^2 is unity.) Frequently, the amplitude distribution of a speech signal can be approximated by the PDF given in Fig. 2.3.[5] For signals such as speech signals, where the r.m.s.

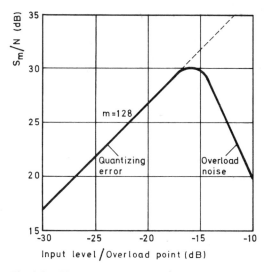

Fig. 2.2 Signal-to-noise ratio S/N as a function of input level in a uniform quantizer

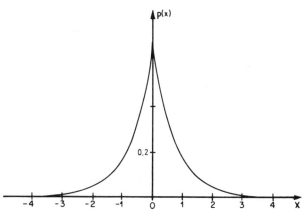

Fig. 2.3 Laplacian probability density function

value is about 13 dB less than their maximum value, the S/N ratio is smaller than that of a sine wave by about 11 dB.[1]

2.2 Codes for Analog-to-Digital Conversion

Coding relates the amplitude levels of a quantized signal to an alphabet system. In digital transmission there are only two symbols, 0 and 1, which are combined with words for establishing different messages to be stored or transmitted.

As mentioned in Chapter 1, the function of a converter is to assign an analog value to one of a finite number of digital values. These values can be established by well-defined rules from the set of symbols representing the code words. These rules differ for various codes of equal length. In Chapter 1 the code was established by the coefficients $b_0 \ldots, b_{n-1}$ of the binary numbering system used to represent a signal amplitude s. For convenience, this notation is given again in the form

$$s = b_{n-1} \cdot 2^{-1} + b_{n-2} \cdot 2^{-2} + \ldots + b_0 \cdot 2^{-n}$$

This method of coding is appropriate for converters because the range taken by s is interpreted as a fraction of full scale. For instance, the fractional code word 01001 has a value of $(0 \times 0.5) + (1 \times 0.25) + (0 \times 0.125) + (0 \times 0.0625) + (1 \times 0.03125)$, or 0.28125 of the full-scale value. If all the bits are ones, the result is not full scale but $(1 - 2^{-n}) \times$ full scale. Thus for an 8-bit D/A converter with a full-scale output range of 10 V and all its digital inputs being one (i.e. 1111 1111), the actual analog output value is

$$(1 - 2^{-8}) \cdot 10 \text{ V} \approx 9.9609 \text{ V}$$

The quantization interval Q, or LSB size, is full scale divided by 2^n; in the case above it is about 39 mV.

Depending on the structure of the A/D or D/A converter, it may be advantageous to use a certain code. We will elaborate on this in connection with the conversion process itself. Here some selected codes are presented to introduce the subject.[6-8]

There are unipolar and bipolar codes, depending on the sign of the measuring range. Most common unipolar codes are straight binary, complementary binary, binary-coded decimal (BCD), and complementary BCD. These are listed in Table 2.1 for an 8-bit word length and two widely used full-scale ranges of +10 V and +5 V, respectively. The most common bipolar codes are given in Table 2.2 for the same word length and full-scale ranges as given in Table 2.1.

In this context it is also useful to remember the two definitions in use for digital circuitry. If the 'High' level (i.e. +5 V for TTL) is assigned to the binary ONE, it is called positive logic or positive true logic. Consequently, the 'Low' level (i.e. 0 V for TTL) is assigned to ZERO. In negative true logic the reverse applies, i.e. a 'High' level is related to ZERO (+5 V in TTL) and a 'Low' level to ONE (i.e. 0 V in TTL).

The straight binary code consists of the coefficients $b_{n-1} \ldots b_0$, with the most significant bit (MSB) at the first position and b_0 at the last. Complementary binary is obtained by inverting bit by bit. Thus, complementary binary in positive true logic is identical to straight binary in the negative true logic. The negative true specification is used in minicomputer bus systems for input/output.

In converters used in digital voltmeters, for instance, a common code is binary coded decimal (BCD). Four bits with the weights 8421 are needed to represent one decimal figure from 0 to 9. There are other decimal codes using other weights such as 2421, 5421, 5311, or excess-three. All of them are relatively inefficient

Table 2.1   Unipolar codes for an 8-bit word length

Scale	FS = 10 V	FS = 5V	Straight binary	Compl. binary
+FS − 1 Q	9.961 V	4.981 V	1111 1111	0000 0000
+7/8 FS	8.750 V	4.375 V	1110 0000	0001 1111
+3/4 FS	7.500 V	3.750 V	1100 0000	0011 1111
+5/8 FS	6.250 V	3.125 V	1010 0000	0101 1111
+1/2 FS	5.000 V	2.500 V	1000 0000	0111 1111
+3/8 FS	3.750 V	1.875 V	0110 0000	1001 1111
+1/4 FS	2.500 V	1.250 V	0100 0000	1011 1111
+1/8 FS	1.250 V	0.625 V	0010 0000	1101 1111
0 + 1 Q	0.039 V	0.019 V	0000 0001	1111 1110
0	0.000 V	0.000 V	0000 0000	1111 1111

Input level ─────→ A/D conversion ─────→ Output code

Output level ←───── D/A conversion ←───── Input code

+FS − 1 Q	9.9 V	4.45 V	1001 1001	0110 0110
+90/100 FS	9.0 V	4.50 V	1001 0000	0110 1111
+80/100 FS	8.0 V	4.00 V	1000 0000	0111 1111
+70/100 FS	7.0 V	3.50 V	0111 0000	1000 1111
+60/100 FS	6.0 V	3.00 V	0110 0000	1001 1111
+50/100 FS	5.0 V	2.50 V	0101 0000	1010 1111
+40/100 FS	4.0 V	2.00 V	0100 0000	1011 1111
+30/100 FS	3.0 V	1.50 V	0011 0000	1100 1111
+20/100 FS	2.0 V	1.00 V	0010 0000	1101 1111
+10/100 FS	1.0 V	0.50 V	0001 0000	1110 1111
+0 + 1 Q	0.1 V	0.05 V	0000 0001	1111 1110
0	0.0 V	0.00 V	0000 0000	1111 1111
Scale	FS = 10 V	FS = 5V	Binary coded decimal	Complementary BCD

because they need a 4 binary digit capable of representing sixteen values to describe ten values only.

To obtain bipolar operation the positive unipolar range is offset by one half of full scale or by the MSB value, i.e. it is shifted down in order to obtain negative values. If the all-zero code word is assigned to negative full scale and increasing (i.e. less negative) analog values are related to increasing binary code words, the offset binary code is obtained. If the maximum positive voltage, for instance, 5 V (compare Table 2.1), is maintained, the size of the quantizing step is doubled. This is assumed in Table 2.2.

The two's complement code is the same as offset binary except for the first bit, which is inverted (sign bit). This code is useful in arithmetic operations. In it the sum of a positive and a negative number equal in magnitude, is zero plus a carry.

Table 2.2 Bipolar codes for an 8-bit word length

Scale	FS = ±5 V	Offset binary	Two's compl.	One's compl.	Sign + mag.
+FS − 1 Q	+4.961 V	1111 1111	0111 1111	0111 1111	1111 1111
+3/4 FS	+3.750 V	1110 0000	0110 0000	0110 0000	1110 0000
+1/2 FS	+2.500 V	1100 0000	0100 0000	0100 0000	1100 0000
+1/4 FS	+1.250 V	1010 0000	0010 0000	0010 0000	1010 0000
0	0.000 V	1000 0000	0000 0000	0000 0000[a] / 1111 1111[a]	1000 0000[a] / 0000 0000[a]
−1/4 FS	−1.250 V	0110 0000	1110 0000	1101 1111	0010 0000
−1/2 FS	−2.500 V	0100 0000	1100 0000	1011 1111	0100 0000
−3/4 FS	−3.750 V	0010 0000	1010 0000	1001 1111	0110 0000
−FS + 1 Q	−4.961 V	0000 0001	1000 0001	1000 0000	0111 1111
−FS	−5.000 V	0000 0000	1000 0000	−	−

[a]One's complement and Sign + magnitude binary have two words for zero; the upper is called 0+, the lower, 0−.

For positive values the one's complement code is the same as the two's complement code. For negative values the bit positions of the positive values, equal in magnitude, have to be inverted. Finally, the sign-magnitude binary code equals offset binary for positive values with only the first bit inverted for negative values of equal magnitude. The relationships between bipolar codes are given in Table 2.3.

One's complement and sign-magnitude codes have magnitudes that increase from zero to plus full scale and from zero to minus full scale. Due to their symmetry, both these codes have two code words for zero, their range being one LSB less than for offset-binary and two's complement coding. Coding can also be used to detect and correct errors of transmission or storage. (More on this subject can

Table 2.3 Relations among codes

	Offset binary	Two's complement	One's complement	Sign-magn. binary
+FS	111............11	0 11............11		
>0	... 10 ↑ Increasing values	Invert sign bit of offset binary	Equals two's complement	Equals offset binary
<0	... 10 ... 01	(Sum of positive and negative number of equal magnitude equals zero plus carry)	Invert all positions for positive value	Complement sign bit of equal magnitude positive code
−FS	000............00	1 00............00		

be found in Reference 2.) Error correction is also employed in A/D converters and this will be described in more detail in section 3.3.2.

2.3 Efficient Coding

Providing a number of bits $n = \log^2 m$ as many times as is required by the Nyquist frequency is by no means a very efficient way of communicating the messages represented by the code words. This can be understood by considering the fact that fulfilling the sampling theorem for the highest frequency contained in the signal spectrum means that too much is done for the lower frequencies within the signal spectrum. Most real world signals, such as sounds or images, concentrate their energy at frequencies far below the band edge frequency (see Fig. 2.4). For example, for a telephone signal band-limited to below 4 kHz, sampling at 8 kHz avoids aliasing or spectrum overlap. For a 1-kHz input, four times more samples are taken than needed: three out of four samples are redundant.

Efficient coding means eliminating the redundant amount of information produced by the straightforward A to D conversion process. Here again, only an extract of the significant field of information theory which can be applied to the field of converters will be given. The reader is referred to the literature at the end of this chapter for further information.[9-18]

In the light of information theory, the signal to be converted by the A to D converter originates from an information source that communicates messages to a receiver or information sink. These messages can be separated into various parts as indicated in Fig. 2.5. A vertical line separates the plane into a redundant and a non-redundant part. (A redundant part is that portion of the message which the receiver knows already or which it can predict from transmitted messages.)

A horizontal line separates the plane into a relevant and an irrelevant part. Irrelevant messages are those which the receiver is not interested in (i.e. noise) or which it is not able to detect (inaudible distortions, invisible brightness changes). Efficient coding removes both the redundant and irrelevant part of the messages.

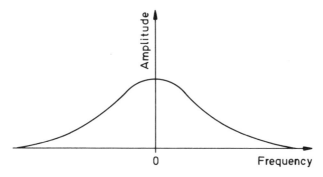

Fig. 2.4 Spectrum of typical 'real world' signals

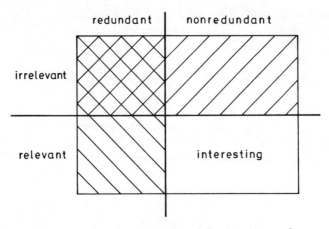

Fig. 2.5 The information plane defined by Shouton[9]

While the redundant part can be completely reconstructed on the receiver side (i.e. the process is reversible), the removal of the irrelevant part represents an irreversible process which draws information from the signal that is lost. The amount of irrelevance must be defined by the receiver and is related to the fidelity of the reproduction of the signal waveform.

2.3.1 Coding of Independent Messages

The messages produced by an A to D converter consist of the code words representing the various amplitude levels. If the signal source has no memory, they are independent of each other. The information I to be derived from the number of levels m is

$$I = \log_2 m$$

The unit of information is the binary digit (= bit). The number equals the number of binary coefficients of value 0,1 needed to represent all m levels in the binary code, as explained in section 2.1. The information rate is determined by $C = f_s \cdot I$, where f_s is the Nyquist frequency.

The average information content of a message called entropy H depends on the probability of occurrence $p(V_i)$ of the different amplitude levels V_i:

$$H = -\sum_{i=1}^{m} p(V_i) \cdot \log_2 p(V_i)$$

Only in the case of equal probability of occurrence $p(V_i) = 1/m$

$$H = -\sum_{i=1}^{m} 1/m \cdot \log_2 (1/m) = -\log_2 (1/m) = \log_2 m = I$$

In general, $p(V_i) \neq 1/m$: then

$$H < I$$

The redundancy R is defined as

$$R = I - H \geq 0$$

Considering a set of $m = 8$ messages A, B, C, D, E, F, G, H, then $I = \log_2 8 = 3$ bit. Let us assume various probabilities of occurrence:

Message:	A	B	C	D	E	F	G	H
Probability:	0.10	0.18	0.40	0.05	0.06	0.10	0.07	0.04

For this distribution $H = 2.55$ bits and $R = 0.45$ bit.

Efficient coding is carried out by allocating short-length code words to frequent messages and longer code words to less frequent ones, as done, for instance, by Shannon et al.[10–12] A simple, but not most efficient (suboptimal), example is given in Fig. 2.6, where the messages are listed first in order of decreasing probability and then combined into pairs of nearly equal occurrence. New combined messages are established that are again arranged into pairs. By assigning the binary figures 0 and 1 to the messages existing at the beginning and at intermediate stages, a coding tree (Fig. 2.6) can be formed where each message has a unique code word.

Considering the word-length and its probability, the average word-length can be calculated to be $L = 2.64$ bits, leaving a redundancy of $R = L - H = 0.11$ bit. The coding procedure is implemented most easily by a read-only memory that stores the table of the reduced redundancy code words and uses the redundant 3-bit words as addresses. The major disadvantage is the unequal length of the code words between 2 and 4 bits requiring a buffer memory to achieve a constant data rate for transmission.

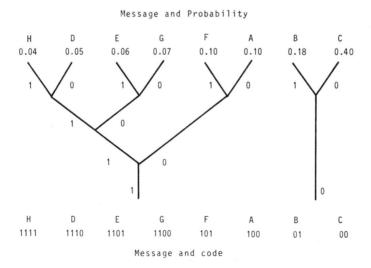

Fig. 2.6 Coding tree

Optimum coding with respect to irrelevance reduction is achieved in the A to D converter if the quantizing characteristic, i.e. the distribution of amplitude levels, is adapted to their probability of occurrence. Taking the minimum mean square error (MMSE) as a fidelity criterion, minimum distortion is achieved by the Max-quantizer[13] if, at a constant word-length, the quantization intervals are chosen such that each contributes an equal amount of quantization noise.

In an article by Gersho[4] zero memory uniform quantizers are compared with those followed by entropy coding, block coding, and non-uniform coding. It is seen that the uniform quantizer achieves a signal-to-noise ratio performance within 7 dB of the best performance attainable with independent samples.

For the more general case of a signal with known statistical properties, the rate distortion function indicates the minimum bit rate needed to maintain an average signal-to-noise ratio for the transmission of that signal.[14]

Another means of reducing the required word-length which is in widespread use is known as 'companding'. This is described in more detail in section 3.6. In this case, quantization becomes more coarse as amplitude increases, yielding a nearly constant signal-to-noise ratio over a wide input signal amplitude range. This is significant for speech and audio signals as well as for instrumentation and control, where the complexity of a coder would become prohibitive if the resolution needed for low-level signals would be linearly extended over several decades.

2.3.2 Predictive Coding

If the signal source has memory, which is the case in most real world signals, the signal samples are not independent of each other, i.e. they are correlated. Any sample to be encoded can be predicted, within certain limits, from its predecessors. If the correlation extends over p previous samples, the source is called a Markoff source of order p. Prediction removes redundancy, i.e. it decorrelates the signal. This can be done in various ways, all of them carrying out a filtering process. When looking at the spectrum of a correlated signal it is seen that it has its energy concentrated on the lower parts of the frequency spectrum (Fig. 2.4) far below the Nyquist frequency.[15] Decorrelated signals making efficient use of the coder, in the sense of equal distribution of code words, have a flat spectrum. So decorrelation means filtering the signal with a filter having the inverse response of the signal spectrum (Fig. 2.7). This can be done in the frequency domain in front of the encoder, which is known as 'pre-emphasis'.

Here we are concerned with performing the filtering function digitally in the form of a predictive coder which is a digital high-pass filter. Consider the first-order predictive coder (Fig. 2.8), which takes the signal s_n to predict the next sample $\hat{s}_n$ by weighting s_n by the predictor coefficient $a < 1$ and delaying it for one sampling interval T_s; $e_n = s_n - \hat{s}_n$ is the prediction error which is transmitted. The frequency response is

$$E(j\omega)/S(j\omega) = 1 - H_0(j\omega) = 1 - a \cdot e^{(-j\omega T)}$$

the magnitude of which is shown in Fig. 2.9 (H_0 is the pulse response). It is similar to that of Fig. 2.7, shown to decorrelate a redundant signal.

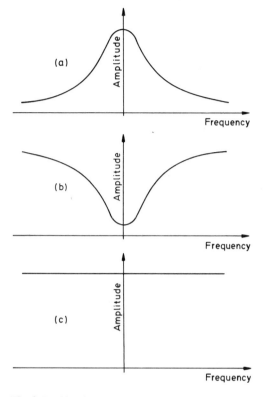

Fig. 2.7 Shaping the input spectrum by filtering. (a) Spectrum of signal; (b) filter response; (c) spectrum at the input of an ADC (log scale for amplitudes)

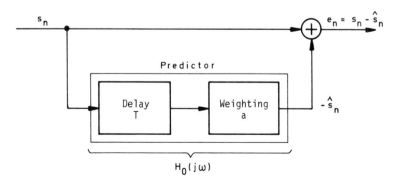

Fig. 2.8 Basic predictive structure

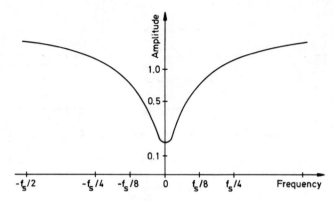

Fig. 2.9 Frequency response of the structure shown in Fig. 2.8 ($a = 0.8$, log scale for amplitudes)

In order to achieve complete decorrelation the predictor has to calculate its prediction value $\hat{s}_n$ from all the previous p samples representing the memory of the source. The weighting coefficients a_i are to be derived from the autocorrelation function.[16] The process of decorrelation itself does not remove redundancy; the entropy of the difference signal e_n equals that of s_n. However, the sample values e_n have a high probability at low values and a small probability at high magnitudes (Fig. 2.10).

The redundancy could be removed by a suitable code. In practice, however, a quantizer is chosen with a much smaller number of quantizing levels, i.e. it considers the small number of large magnitude differences to be irrelevant. This can be justified on the basis of the signal-to-noise ratio. In other words, the power of the prediction error e_n is much smaller than that of the original signal s_n. So encoding e_n with the same quantizing noise power as s_n needs a much smaller

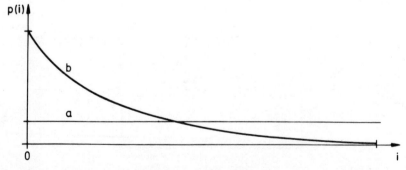

Fig. 2.10 Probability density function at the input (a) and at the output (b) of a decorrelator (schematic)

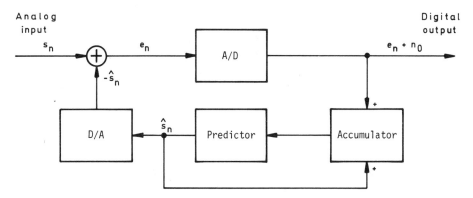

Fig. 2.11 Predictive encoder (n_0 represents the quantization noise)

number of quantizing levels m'. The prediction gain then can be expressed as

$$G_p = \log_2 m/m' \text{ bits}$$

or

$$G_p/\text{dB} = 6 \cdot G_p/\text{bits}$$

G_p depends on the signal and the order p of the predictor. To give two examples: for video signals G_p can be as high as 4 to 5 bits per sample when starting from an 8-bit linear encoding of a standard NTSC or PAL signal. For telephony speech it can also assume 3 to 4 bits when starting from a companded 8-bit PCM sequence at a 8 kHz sampling rate.

A predictive encoder is established by introducing a predictor in the feedback loop (Fig. 2.11). The predictor calculates its estimation $\hat{s}_n$ from the quantized output values of the ADC, which gives two distinct advantages. First, the error of the quantization process can be corrected by the next sample; second, the receiver or decoder (Fig. 2.12) can ensure that the accumulation of the differences e_n

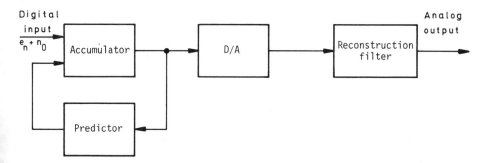

Fig. 2.12 Predictive decoder

approaches the signal within 9/2. If a predictive coder is established it enables observation of the signal used for reconstruction at the output of the accumulator in the feedback loop, which is the input to the predictor. A decoder which has to follow a predictive encoder is shown in Fig. 2.12. For reasons of loop stability and robustness with respect to transmission errors, the sum of the magnitudes of the prediction coefficients should stay below 1.[16]

The prediction gain reduces the number of bits needed to transmit the signal for a prescribed fidelity. It also reduces the number of quantizing levels of the ADC in the feedback loop of Fig. 2.11 because the error signal e_n has a correspondingly reduced dynamic range. These savings must be compared with the additional cost of the predictor, accumulator, and D/A-converter.

Another technique for successful reduction of the number of quantizing levels within the ADC has been developed in interpolative coders, which are described in section 3.4.4. Using this technique, the errors by coarse quantization are measured several times within one sampling interval, i.e. the signal is oversampled and added up by integration. The number of times the error has to be added up to cross the next quantizing level is a measure of the fraction that the input signal value takes between two adjacent quantizing levels.

When sampling higher than the Nyquist frequency the quantizing noise covers a correspondingly wide frequency band. This noise can be shaped by a filter to fall substantially beyond the signal frequency band.[17] In this way, the signal-to-noise ratio is increased within the signal band. This gain, due to oversampling and noise shaping, is compared with the gain by differential encoding for both analog and digital predictors, including quantizing level inaccuracies of the ADC and DAC.[18] It is claimed that oversampled DPCM techniques are not superior to simply oversampled A/D converters. However, the signal considered is a frequency-division multiplexed (FDM) groupband signal of twelve voice channels between 60 kHz and 108 kHz, which does not seem to be especially suitable to DPCM coding.

References

1. Cattermole,, K. W., *Principles of Pulse Code Modulation*, Illife Books, London (1969).
2. Taub, H., and Schilling, D. L., *Principles of Communication Systems*, McGraw-Hill, Kogakusha (1971).
3. Hoelzler, E., and Holzwarth, H., *Pulstechnik*, Vol. 1, 2nd edn, Springer, Berlin/Heidelberg (1982).
4. Gersho, A., 'Principles of Quantization', *IEEE Transactions* **CAS-25**, 427–36, July (1978).
5. Rabiner, L. R., and Schafer, R. W., *Digital Processing of Speech Signals*, Prentice-Hall, New Jersey (1978).
6. Bruck, D. B., *Data Conversion Handbook*, Hybrid Systems Corpn, Burlington, Mass. (1974), Chapter 1.
7. Hnatek, E. R., *A User's Handbook of D/A and A/D Converters*, Wiley, New York (1976), Chapter 3.
8. Zuch, E., 'Know your Converter Codes', *Electronic Design* **22**, 130–35, October (1974).

9. Shouton, J. F., 'Nachricht und Signal', *Nachrichtentechnische Fachberichte*, 6, II.1–II.2 (1957).
10. Shannon, C. E., 'A Mathematical Theory of Communication', *BSTJ* 27, 379–423, 623–56 (1948).
11. Fano, R. M., *The Transmission of Information*, MIT Report 65 (1949).
12. Huffman, D. A., 'A Method for the Construction of Minimum Redundancy Codes', *Proceedings of the IRE* 40, 1098–1101 (1952).
13. Max, J., 'Quantizing for Minimum Distortion', *IRE Transactions* IT-6, 7–12, March (1960).
14. Davisson, L. D., 'Rate Distortion Theory and Application', *Proc. IEEE* 60, 800–8, July (1972).
15. Karkowski, R. J., 'Predictive Coding Improves ADC Performance', *Electronic Design News* 24, 137–43, October 5 (1979).
16. Arp, H., 'Exact Models for Predicting Picture Signals concerning Adaptive DPCM-Transmission', Paper presented at the 1974 Picture Coding Symposium, Goslar, West Germany, 26 August 1974.
17. Tewkesbury, S. K., and Hallock, R. W., 'Oversampled, Linear Predictive and Noise-Shaping Coders of Order $N > 1$', *IEEE Transactions* CAS-25, 436–47, July (1978).
18. Cyganski, D., and Krikelis, N. J., 'Optimum Design, Performance Evaluation, and Inherent Limitations of DPCM Encoders', *IEEE Transactions* CAS-25, 448–60, July (1978).

CHAPTER 3
Analog-to-Digital Conversion Techniques

This chapter represents the core of the book. It deals with the wide variety of methods, structures, and algorithms that have been developed for A/D conversion by designers in order to meet the criteria of performance and cost.

3.1 Classification of the Methods

A/D converters have evolved historically from various disciplines with little relationship to each other, such as communications and instrumentation. Therefore it is difficult to establish today a classification system based on criteria that are mutually exclusive[1] and which, in addition, makes sense only at a certain level of common understanding between designers and users in various fields.

The classification scheme used throughout this chapter is based on converter architectures, i.e. structural parameters that manifest themselves within the circuit block diagram of the converter, a representation with which most engineers dealing with converters are familiar. So the three major categories are parallel, cascade, and feedback converters, which conform with the popular parallel, sequential/parallel, and pure sequential categories.

Although differing only in their implementation, indirect converters deserve a separate section of this book, as do others grouped under miscellaneous converters, especially those for special codes and/or applications. Microcomputers are not only increasing the scope of A/D converter applications, they also add to the freedom of the designer of ADCs, and therefore merit a section in this chapter.

The summary at the end of the chapter attempts to relate the classification chosen to others already established.

3.2 The Parallel (Direct) Converter Structure

This principle is explained by using a mechanical analogy, i.e. the process of measuring the unknown length, X, of a stick (Fig. 3.1), using a reference scale of length $m \times Q$, where Q is the size of the quantization interval (quantum). Since all values are simultaneously present on the scale, the measurement is accomplished directly in a single step, leading to the term 'word-at-a-time', i.e. the direct or simultaneous method. Electrical realization needs $(m-1)$ comparators and references spaced by Q.

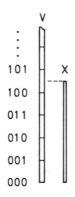

Fig. 3.1 Direct measurement of a length X by comparison with a reference V

The major merit of this technique is its high speed; its major drawback is its high component count. This may be compensated to a certain extent by operating the converter without a sample-and-hold circuit in front of it, in the so-called 'flash' mode, where the sampling operation is accomplished by strobing or gating digitally (see also Chapter 5).

A possible common error of this method should be mentioned. Consider again the so-called position encoder. As shown in Fig. 3.2, a ruler has n-longitudinal segments corresponding to the length $n = 4$ of the code word. The black areas are connected to a voltage representing a logic '1' within the code, the white areas lead to logic '0' (no voltage). The position is sensed by four brushes (or sensors) directly giving the position in binary code form. The bits of the binary code are found independently. Therefore if the sensor-set is slightly skewed (shown dashed in Fig. 3.2), the reading will be erroneous at the transition of the segments. In the worst case, when a change in the MSB takes place, the resulting error reaches one half of full scale. Other codes such as the Gray code overcome this deficiency by only one digit change between adjacent positions, as explained in the next section.

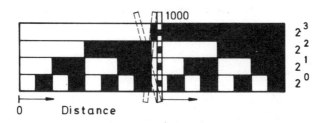

Fig. 3.2 The position encoder

3.2.1 The Cathode Ray Encoding Tube

The cathode ray encoding tube represents an early electrical realization of the parallel method, and it was designed for high-speed applications of about 12 MHz word rate.[2] The structure of the coding tube as illustrated in Fig. 3.3(a) is similar to that of the conventional cathode ray tube. However, the electron beam takes the shape of a thin horizontal blade (ribbon beam). Therefore no horizontal deflection of the beam takes place. The target consists of a coded aperture plate and a collector plate. The first plate has openings according to the code as shown in Fig. 3.3(b). The other carries a detecting segment for each bit position of the code and is called the detector—or collector plate. The output is delivered in parallel form.

Upon applying the input analog voltage to the vertical deflecting plates the electron beam scans across the aperture plate, causing an output on the unshielded detector segments. As mentioned in the previous section, to prevent errors the Gray code is used, since only one bit changes state for consecutive code values.

As shown in Table 3.1, the MSB is the same for both codes, the Gray code and the straight binary code. The other bits of the Gray code are obtained simply by reflecting the bits of a certain column at the zero/one transition of the next more significant bit. Therefore this code is also referred to as the 'reflected binary code' or, alternatively, the 'unit-increment code'. Table 3.1 shows the binary and Gray codes. However, the Gray code is not suitable for performing mathematical operations since 'weights' can no longer be assigned to bit positions.

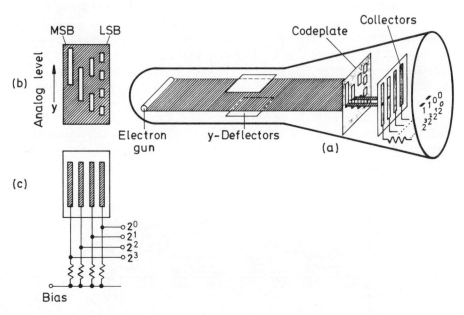

Fig. 3.3 The cathode ray encoding tube. (a) Total view; (b) aperture code plate; (c) collector plate

Table 3.1 Comparing binary and Gray coding

Decimal	Binary	Gray
15	1111	1000
14	1110	1001
13	1101	1011
12	1100	1010
11	1011	1110
10	1010	1111
9	1001	1101
8	1000	1100
7	0111	0100
6	0110	0101
5	0101	0111
4	0100	0110
3	0011	0010
2	0010	0011
1	0001	0001
0	0000	0000

To convert the Gray code into binary code, exclusive-OR gates are used, as shown in Fig. 3.5 for a 4-bit word. The exclusive-OR gate, whose symbol truth table and structure are given in Fig. 3.4, realizes the logic function

$$E_0 = (E_1 \cdot \bar{E}_2) + (\bar{E}_1 \cdot E_2)$$

Recently the coding tube has been upgraded in speed to a word rate of 200 MHz at 8-bit[3] by introducing an electron-bombarded semiconductor target (EBS). The target consists of an array of eight long planar diodes that are made conductive if the electron beam hits the surface at a location which is not masked by metal. The silicon diodes exhibit sufficient gain even at the high rate of change of the LSB, which is 128 times that of the applied signal frequency for a full-scale sine wave.

For this device a rigid test procedure has been established[4] based on the S/N-ratio achieved at a specified frequency and amplitude. According to this high

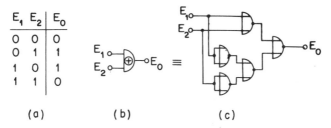

Fig. 3.4 The exclusive OR-gate. (a) Truth table; (b) symbol; (c) decomposition into NOR gates

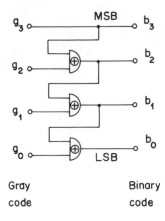

Fig. 3.5 Circuit for converting 4-bit Gray code to 4-bit binary code

standard, the tube is capable of encoding a 20 MHz half-scale sine wave at 6-bit accuracy, which is considered to be well ahead of all state-of-the-art solid state converters (this is described in Chapter 6 in connection with measurements).

3.2.2 Parallel Converters with Voltage Comparison

An electrical ruler calibrated in steps corresponding to the size Q of the quantum interval can be implemented by connecting a series string of equal value resistors R to a reference voltage V_{ref}. As shown in Fig. 3.6, the voltages developed at the taps of the resistor chain supply reference voltages to a bank of comparators, the second inputs of which are connected in parallel to the unknown analog voltage V_x via a buffer amplifier. The outputs of these comparators whose reference voltages are smaller than V_x are high, i.e. '1'. Other outputs assume the state '0'. At a certain rate dictated by the sampling frequency, these states are transferred to a corresponding number of latches, where they are stored. In a decoding logic the digital output is then obtained in the required code. In case of binary coding of word-length n, the required number of comparators is

$$N = m - 1 = 2^n - 1$$

In practice, the number will be 2^n to establish overflow control.

The amount of conversion hardware associated with this method is reflected in the large number of comparators and latches that is required. In addition, there is a need for a high-quality buffer amplifier that is able to feed, both statically and dynamically, this large number of comparators in parallel. In addition, these comparators must always have an increased sensitivity for a fixed full-scale range, or, alternatively, for a certain sensitivity a larger common mode input range is needed.

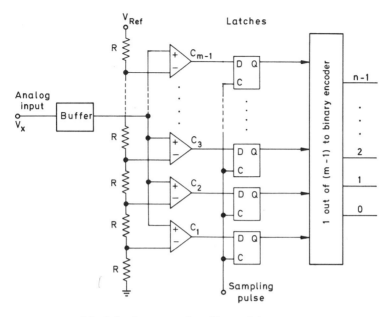

Fig. 3.6 Structure of a n-bit parallel converter

If the converter is preceded by a sample-and-hold circuit (S/H, see section 5.2) a sufficient delay must be provided between the S/H gating pulse and the strobing pulse to the latches in order to ensure the settling of the output state of the comparators before latching. Due to the monotonic increase of the reference voltages connected to the comparators (Fig. 3.6), the comparators' outputs and the resulting bits are not determined independently. Therefore the serious error of the position encoder does not occur.

However, if the circuit is operated without an S/H in front, as described in section 5.2.4, so that sampling is done in the digital circuit part, special precautions must be taken, as described below, to compensate for the variation of the comparators' response times.

Early parallel converters used hybrid integration where, for example, a chip carrying eight monolithic comparators is connected to another chip containing eight gates on a thick-film substrate.[5] Sixteen substrates establish a 7-bit A/D converter operating at a sampling rate of 30 MHz. The decoding logic is realized by the so-called *V*oltage *A*ddressable *R*ead *O*nly *M*emory (VAROM), having 128-word capacity at 7-bit length. The ROM diodes were based on the high-speed Silicon-On-Sapphire (SOS) technology. Figure 3.7 shows how the addresses of the memory locations are obtained. Each two inputs of the NOR-gates are connected to the Q_i and the Q_{i+1} outputs of the *i*th and $(i+1)$th comparators, respectively. In this way a logical 1 is obtained at the output of the *i*th gate only if $Q_i = 0$ and $Q_{i+1} = 0$. Moreover, the sampling function is included in the comparators (see section 5.3).

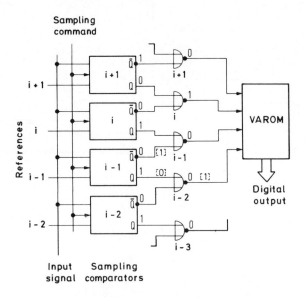

Fig. 3.7 Encoding in a parallel A/D converter by a voltage addressable read only memory (VAROM[5])

Different response times of the strobed comparators can cause errors. Such a case is indicated in dashed brackets in Fig. 3.7. Comparator $(i-1)$ had not yet settled when it was strobed. Therefore gates $(i-2)$ and i show a ONE at their Q–output. In a chain of comparators with a logic '1' at their output only the 'leading' one should be encoded to avoid serious errors. Consequently, in an all-parallel ADC operating without an S/H priority encoding of the comparators' outputs must take place.

This is done in another hybrid 6-bit flash converter which proved to be very resistant to dynamic errors. It was built by Tietze[6] using standard ICs to operate at a word-rate of 20 MHz. Schottky-TTL comparators (NE521) and standard TTL-elements of the 74-series for storage were used. The design is similar to the basic structure given in Fig. 3.6, except for the coding logic. Encoding is implemented by using eight input priority decoders (TTL-type 74148). As shown in Fig. 3.8, eight decoders (PDA–PDH) are required for the 64 outputs of the latches that sample the output of the comparators. The first three outputs of each PD represent the binary coded address of the highest position input having a logic '1' (position corresponds to amplitude). At the fourth output of each decoder the state 1 is obtained if any input of the respective decoder is high. All these number 4 outputs are then connected to another priority decoder, PDI, that delivers at its outputs (I1, I2, and I3) the three most significant bits of the output word. These bits provide the address of the multiplexers MUX1, MUX2, and MUX3 which give the three lower significant bits. For this purpose, the first output (A1, B1, ...

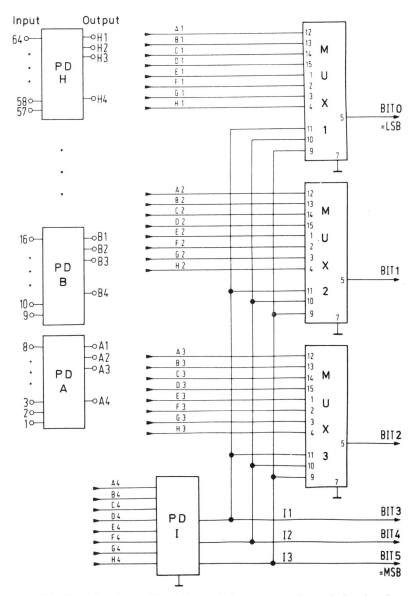

Fig. 3.8 Encoding logic of a 6-bit parallel converter using priority decoders (PD)

H1) of each decoder is connected to the eight inputs of the first multiplexer M1, the second outputs to M2, and the third outputs to M3, respectively. Measurements on this converter have shown its superior performance.

Monolithic circuit technology provides both high speed and reasonable cost when the complete parallel converter can be placed on one chip or a sub-unit can be extended to higher resolution by having additional units in one single integrated circuit package. A monolithic 5-bit building block based on NMOS circuit technology with 7 μm minimum line width is intended as a building block for higher resolution. Its coding logic is a fast read-only memory. The circuit operates up to a clock rate of 20 MHz.[7] A 6-bit expandable unit based on CMOS/SOS (SOS = silicon on sapphire) works up to 40 MHz clock rates and up to 15 MHz in the flash mode.[8] Both monolithic building blocks have a low power dissipation due to the MOS technology they are using, the quoted parameters being 550 mW for the NMOS and 40 mW for the CMOS version, respectively.

Bipolar devices are capable of achieving even higher speeds: a monolithic 6-bit parallel converter based on ECL circuitry in a standard bipolar technology ($f_t = $ 1, 5 GHz) runs at clock rates up to 150 MHz.[9] Its accuracy is sufficient to be extended to 8-bit resolution, and its power dissipation of 450 mW is very low for bipolar devices.

A standard for monolithic A/D converters has been established by an 8-bit monolithic bipolar converter capable of 35 MHz rate in the flash mode.[10] This consists of 255 strobed comparators, clock buffers, combining logic, and an output buffer register, all on one chip. Power dissipation amounts to 2.5 W. The input capacitance to be driven by a buffer amplifier is 300 pF. Recently, a new design with line dimensions of the bipolar process reduced to 1 μm can achieve 8-bit and can be operated up to a 75 MHz word-rate.[11] It is reputed to handle a 20 MHz bandwidth signal at 2 W power dissipation.

A modification of the parallel converter for 8 bits resolution at 30 MHz word-rate uses a two-step encoding procedure.[98] A resistor string provides 255 reference voltages. In the first step, the signal range is quantized into sixteen levels by fifteen coarse comparators. The second step uses the output of the coarse encoding to switch the supply current to fifteen primary comparator stages for further subdivision of the coarse intervals. The primary comparator outputs are fed to fifteen fine comparators that are shared by the primary comparators. The primary comparators are said to replace the function of the DAC needed in a two-stage sequential/parallel design.

It should be noted that clock-rate, word-rate, or flash mode repetition frequency should not be confused with the highest full-scale range sine wave that can be coded at the same resolution specified for low frequencies. In some of the references, 3-dB bandwidth of the comparators are quoted that are similar to or below the highest clock rate. This indicates that the frequency for 1/2 LSB degradation is far below the clock rate. To obtain comparable performance values it is important to agree on standard test procedures for dynamic performance testing (see Chapter 6).

3.2.3 A Parallel Converter with Current Comparison

A considerable part of the cost of a parallel converter is due to the pre-amplifier that feeds, in parallel, a large number of voltage-comparators. At high frequencies the input capacitances of the comparators present a considerable capacitive load to the amplifier, limiting the dynamic frequency response more than any other effect.

To avoid this difficulty, current- rather than voltage-comparison is suggested.[12] A simplified circuit diagram of a parallel converter using current comparison for $m = 8$ is illustrated in Fig. 3.9. The currents flowing in the different branches are given by:

$$i_1 = i_2 + I_{0/8} = i_x - I_{0/8}$$
$$i_2 = i_4 + I_{0/4} = i_x - I_{0/4}$$
$$i_3 = i_2 - I_{0/8} = i_x - 3I_{0/8}$$
$$i_4 = \quad\quad\quad = i_x - I_{0/2}$$
$$i_5 = i_6 + I_{0/8} = i_x - 5I_{0/8}$$
$$i_6 = i_4 - I_{0/4} = i_x - 3I_{0/4}$$
$$i_7 = i_6 - I_{0/8} = i_x - 7I_{0/8}$$

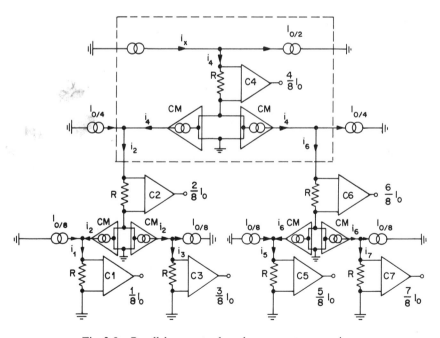

Fig. 3.9 Parallel converter based on current comparison

The current I_0 represents the maximum analog value. In the comparators $C_1 \ldots C_7$ the voltages developed across equal resistors R due to the currents $i_1 \ldots i_7$ are compared with ground potential. The comparators' outputs represent multiples of the increment one-eighth of the full scale. Current mirrors (CM) having a current gain of 2 are used wherever the currents i_2, i_4, and i_6 are split, thereby eliminating the need for high-sensitivity comparators. In an experimental set-up a word repetition frequency of 37 MHz at 4-bit resolution is obtained.

3.3 Cascaded Converters

3.3.1 Introduction

A feature common to all cascaded converters is that the signal to be digitized propagates, as shown in Fig. 3.10, through a number of cascaded stages. Each stage usually delivers one or more bits such that the n-bit output word, obtained from a chain of i stages, is given by

$$n = n_1 + n_2 + \ldots + n_i$$

where $n_1, n_2, \ldots, n_i$ are the number of bits produced in the first, second, and ith stage, respectively.

One characteristic common to all cascaded-architecture converters is the fact that the total range is divided into 2^{n_1} coarse intervals in the first block, and the input voltage is measured according to this coarse scale yielding the most significant bits. An analog voltage, equivalent to the coarse digital output, is then subtracted from the original input. The remaining difference signal is fed to the next stage. The magnitude of the remainder signal is smaller than one coarse interval. Therefore a finer scale has to be provided only for one sub-range, saving the references for all other $2^{n_1} - 1$ coarse intervals, which is the major advantage of this method.

Best[13] calls this technique 'post-subtractive', because the procedure of comparison first and subtraction afterwards is repeated in the subsequent stages, where the intervals become increasingly finer until the final resolution is reached and the remaining bits are available. Since this method saves references at the expense of more time it allows flexibility for an optimum trade-off between speed and hardware complexity.

Depending on which functional characteristic is intended to be emphasized, converters of this type are called also 'sub-ranging', 'propagation', or 'ripple through' converters. Since each of the building blocks frequently includes a parallel converter, they are also known as 'sequential–parallel A/D converters'.

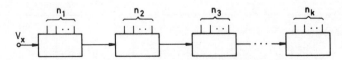

Fig. 3.10   Structure of a cascaded converter

3.3.2 Sequential–Parallel Converters as an Extension of Direct Converters

Consider again the mechanical analogy of measuring the length X of a stick. Instead of performing the measurement in one step, with a large number of comparators and references, one can do it in two or more steps, thereby saving some of the references. For illustration, let us assume two steps. If m_1 coarse intervals covering the whole range were allocated to the first block and m_2 fine intervals to the second, then an overall resolution corresponding to $m = m_1 \cdot m_2$ is obtained. At the same time, the number of the required references N and/or comparators reduces accordingly to

$$N = (m_1 - 1) + (m_2 - 1)$$

Among all possible combinations of m_1 and m_2, the combination that yields the minimum N for a given m is of major interest; this is obtained for $m_1 = m_2$. As an example, for an 8-bit converter where $m = 2^8 = 256$ the number of references in a two-step sequential–parallel design is minimum (i.e. $N = 30$) for $m_1 = m_2 = 16$.

To generalize this consideration, assume that i stages were used. Then m is given by

$$m = m_1 \cdot m_2 \ldots m_i$$

and the corresponding number of references is

$$N = \sum_{k=1}^{i} (m_k - 1)$$

The quantum size in the first stage is equal to m/m_1, in the second it is $m/m_1 \cdot 1/m_2$, and so on. In the ith stage it equals the smallest quantization interval $q = 1$. As stated earlier, N is minimum if the number of intervals is equal for all stages or quantization steps, i.e. when

$$m_1 = m_2 = \ldots = m_i = \sqrt[i]{m}$$

In this case, the number of the necessary references becomes

$$N = i(\sqrt[i]{m} - 1)$$

Usually, the root $\sqrt[i]{m}$ is no integer; hence it is convenient to put

$$m = 2^n \text{ where } n = n_1 + n_2 + n_3 + \ldots + n_i$$

Thus to have a minimum number of references, the condition

$$\sum_{k=1}^{i} n_k \stackrel{!}{=} \text{minimum}$$

must be satisfied. If, for an example of $m = 256$, $i = 4$ is chosen first, then we get

$$m_1 = m_2 = m_3 = m_4 = \sqrt[4]{256} = 4$$

The required number of references for the comparators is $N = 4(4 - 1) = 12$. If $i = 8$ is suggested instead, then all m_k become $m_k = \sqrt[8]{256} = 2$. This means that the quantum size is reduced from one stage to the next by a factor of 2. In this specific case, $i = n = \log_2 m$, i.e. the number of stages is equal to the word-length. This coding method is analogous to the well-known weighting process, and is

Table 3.2 Required number of references for different numbers of stages in an 8-bit cascaded converter

No. of stages i	No. of quantization intervals of blocks	Minimum integer No. of references N	Figure of merit $N \cdot i$
1	256	255	255
2	16, 16	30	60
3	8, 8, 4	17	51
4	4, 4, 4, 4	12	48
5	4, 4, 4, 2, 2	11	55
6	4, 4, 2, 2, 2, 2	10	60
7	4, 2, 2, 2, 2, 2, 2	9	63
8	2, 2, 2, 2, 2, 2, 2, 2	8	64

therefore also called the 'put and take' technique. This can be realized as a post- or a pre-substractive, i.e. a cascade (feedforward) or feedback, structure. The 'number of references' is used as a synonym for the number of comparators or references (see the following sections for more details).

A similar extension is also applicable to the technique of coding by counting (see section 3.4.2). Both extensions meet producing a successive approximation converter that discovers one bit per step. Table 3.2 summarizes the possible solutions for a sequential–parallel converter with $m = 256$. Taking $N \cdot i$ as a figure of merit for overall economy (hardware and time), there is a flat minimum at $i = 4$. For $i > 4$ the number of stages increases faster than the number of references decreases.

A design showing the typical architecture of a sequential–parallel structure for an 8-bit converter with two stages of 4 bit each is given in Fig. 3.11 as an example with more realistic details than in Fig. 3.10.[14] This consists of a first chain of $2^4 - 1 = 15$ strobed comparators followed by the encoding logic for the four most significant bits. These four bits are converted back to analog by a 4-bit D/A converter of 8-bit accuracy, the output of which is subtracted from the analog sample-and-hold output to form a residue which is fed to a second chain of $2^4 - 1 = 15$ strobed comparators and an encoding logic to derive the four least significant bits of the 8-bit word.

Note that in the second block no D/A converter is needed, which would be the case if further extension was intended. The structure lends itself readily to a modular hybrid implementation because two 4-bit parallel converters and a 4-bit D/A converter occur as building blocks. A number of video converters have been based on this structure[15,16] with a bandwidth of up to 5 MHz at sampling rates up to 20 MHz.

As an early landmark in high-speed converters the 6-bit cascaded converter outlined in 1963 by Schindler[17] will be briefly described. This operates at a sampling rate of 50 MHz.

As shown in Fig. 3.12, the converter consists of three blocks cascaded in time due to the delay lines of 6 ns and 12 ns, respectively, in front of their inputs. Each

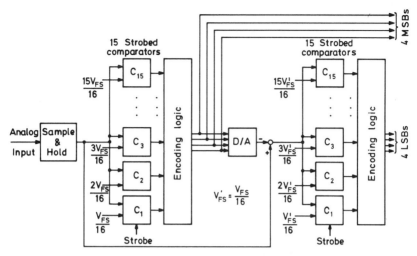

Fig. 3.11 8-bit sequential–parallel converter consisting of two stages of 4-bit each in cascade

block delivers 2 bits in a time of 6.6 ns. The analog input is applied parallel to the stages. As shown in Fig. 3.13, each of the nine channels consists of a buffer amplifier (isolation stage), a high-speed tunnel-diode threshold detector, a memory tunnel-diode, and a current switch representing a quantum interval according to the weight of coarse quantization. The subtraction of coarse intervals is carried out by feeding the current switch outputs back to the inputs of the blocks that

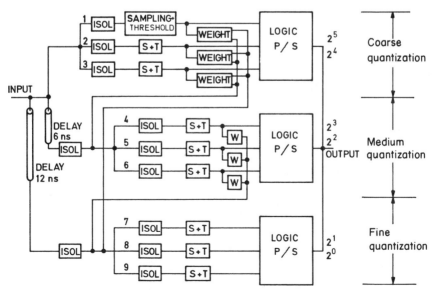

Fig. 3.12 Block diagram of a cascaded parallel converter[17]

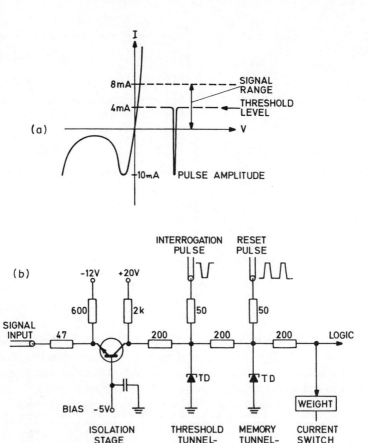

Fig. 3.13 Details of one channel of Fig. 3.12. (a) Current levels in the tunnel diode comparator (b) isolation, threshold, and memory circuit

follow. The required number of comparators is reduced to nine, instead of the 63 that would be necessary for a pure parallel design. For reasons of speed, the decoding logic is placed behind the current switches.

In the last two decades a number of sequential–parallel converters have been built for high-speed applications such as radar, video broadcasting or recording, and nuclear instrumentation. Most of them were made obsolete from the point of view of cost and performance by the monolithic parallel converters now available and described in section 3.2.2. Nevertheless, if ultimate speed and high resolution, or a combination of both, is required to tailor a converter to a specific application a hybrid construction using high-performance comparators, linear wideband amplifiers, and logic chips as building blocks is superior. New challenges come from a high-definition television with a colour subcarrier at 24.3 MHz, digital video recording, digital time base correction, and scan conversion by high-speed full-frame memories, as well as digital television receivers and higher resolution radar. Some examples of these will be given below.

Both Rollenhagen[14] and Saul[18] have designed 8-bit and 10-bit two-stage converters for video speed in hybrid form. A three-stage cascade is used for a 10-bit converter operating a 20 MHz sampling rate, i.e. sufficient for four times colour subcarrier frequency of 17.7 MHz.[19] This consists of two blocks with 4-bit resolution and one 4.5-bit block. The redundancy in the number of comparators is needed to correct for drift and discontinuity effects between adjacent converter blocks (see below). Special attention must be devoted to the interconnection between the stages and the matching of the delays of signals that have to be subtracted from each other, as mentioned elsewhere in this book. This problem is considerably easier to solve in integrated converters with their short internal interconnections, which can be treated as lumped elements instead of transmission lines. Measurements of linearity (33% of one LSB), quantum noise, differential gain (0.4%) and differential phase (0.25°), and operating temperature from 0 to 50°C show the performance achieved. An upgraded three-stage version[20] with 2-bit, 4.5-bit, and 4-bit blocks yielding error correction for a final resolution of 8-bit operates up to 100 MHz sampling frequency.

The technique of increasing accuracy by means of extra bits within the encoder is called 'digital error correction', and is useful for sub-ranging or sequential–parallel converters.[21]

Assume a two-stage sequential–parallel converter with two 4-bit parallel encoders and a 4-bit D/A converter in between, as shown in Fig. 3.11. For simplicity, assume an unrealistic situation, i.e. that the full-scale ranges of the two 4-bit converters are identical, and the same as that of the D/A converter. If there is no error in the first 4-bit ADC, then the residue signal always has to stay within the same quantization interval of the second 4-bit ADC. In this case the second ADC makes no sense at all. However, if the thresholds in the first ADC are too large and the DAC is correct, it will produce too small an output signal, and the residue moves up into the range of the second ADC by as much as the anticipated error of the first ADC. In other words, the error of the first device will be measured and can be used by a digital corrector to produce a correct output. The scheme also works in the more realistic case of only partial overlap of the two 4-bit converters. For instance, a 2-bit overlap produces a 6-bit total output but substantially reduces the accuracy requirements of the first device. Therefore the design can be tailored to the performance of available monolithic building blocks, thus reducing design and manufacturing costs. This digital error correction scheme is the key to a converter of 12-bit resolution working at a maximum sample rate of 5 MHz. In this design, two stages are employed, the first with a 5-bit parallel encoder followed by a 5-bit D/A converter which needs to have full 13-bit accuracy as a reference. The residue is fed to an 8-bit parallel structure. Thus there is one extra bit to correct for the first 5-bit converter deficiencies.[22]

3.3.3 Cascaded 1-Bit per Stage Converters

The simplest conversion hardware in a cascaded structure is obtained if in each stage only one comparison is made. The required number of stages in such a case is then equal to the measuring steps as well as to the length of the output word (n bits), i.e. $i = n = \log_2 m$.

Such a structure was suggested by Smith.[23] As shown in Fig. 3.14, the input voltage V_k to the kth stage of the n similar stages should lie within the unipolar measuring range $A > 0$. A single-bit converter, mainly a comparator whose reference voltage is $A/2$, decides whether the input voltage V_k lies in the upper or the lower half of the full-scale range. It delivers accordingly a single-bit output D_k. This digital output is then converted back to its analog form in a single-bit D/A converter. In a difference amplifier, the converted value is subtracted from the original input V_k and multiplied by a factor of 2 to provide the input V_{k+1} of the following stage. The value of V_{k+1} again lies in the range $0 \leqslant V_{k+1} \leqslant A$. The transfer characteristic of the kth stage, illustrated in Fig. 3.15, is given by $V_{k+1} = 2(V_k - D_k \cdot A/2)$.

The advantage of this method is that identical stages are used allowing a modular construction, and hence extension to higher resolutions is possible. Within each block only simple elements such as comparators, reference voltages, etc. are needed. Also no clock is required, and hence no timing restrictions other than simple delay are to be considered. In addition, the digital output is delivered in parallel form. On the other hand, any amplitude error in the reference voltage is doubled in each subsequent stage. Therefore the accuracy requirements, especially for the first stages, increase with an increased number of bits.

However, a more thorough consideration reveals that timing is to be taken into account. Consider, for example, that the input voltage V_k of the kth stage lies within the range $A/2 < V_k < 3/4A$. Hence the comparator will assume the state '1'. Since V_k during the processing time is applied only to the next stage via the 'feedforward' path and the difference amplifier, then its comparator will assume accordingly the state '1'. After the input V_k has gone through A/D and D/A conversion in the upper branch of the kth stage, the comparator ($k + 1$) switches back to '0'. Thus, to avoid this undesired switching, the comparators should be made

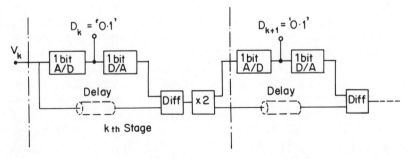

Fig. 3.14 Cascaded structure due to Smith[23]

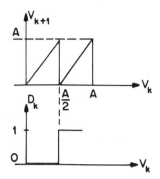

Fig. 3.15 Transfer characteristic of one stage in Fig. 3.14

bistable. Also, a delay element, corresponding to the propagation delay of the A/D and D/A converters of each stage, is introduced in each feedforward path (shown dashed in Fig. 3.14). Furthermore, temporary storage in the form of a sample-and-hold element can be introduced between the stages in order to prevent overshoots of the amplifiers which may otherwise cause coding errors. A most exotic experimental device[24] is reported to encode 5 bits at a 8 GHz rate.

3.3.4 Gray Code Converters

The binary coding obtained by using cascaded architecture is not the only one to represent 2^n values with n bits. Specific modifications lead to the so-called 'Gray Code' (section 3.2.1) that can be transformed with several advantages into a cascaded structure, as we will see.

If the measuring range is bipolar, then a comparison with ground potential will yield the MSB of both the binary and Gray codes. The next bit, as described earlier, could be obtained simply by reflecting the binary bit at the 0/1 transitions of the preceding bit. This coding rule is obtained if the transfer characteristic of each stage in the cascade takes the form

$$V_{out} = -2 \cdot |V_{in}| + V_r,$$

where V_{in} is the input voltage, V_{out} is its output voltage, and V_r is a constant reference corresponding to 1/2 FSR. ($V_{out} = +2|V_{in}| - V_r$ also works but yields complementary 1/0 code values.) This V_{out}/V_{in} characteristic has a V-shape and is usually called the 'folding law' (Fig. 3.16). It is symmetrical to $V_{in} = 0$, to provide the required code reflection. Without any offset, the V-characteristic is equivalent to the characteristic of a full-wave rectifier. By cascading several stages of this type, the required characteristics for Gray coding, shown in Fig. 3.16, can be obtained.

The block diagram of a cascaded converter for Gray coding is shown in Fig. 3.17. The reference voltage of all comparators is ground potential. In addition, no D/A converters and difference amplifiers are needed here as in ordinary cascaded

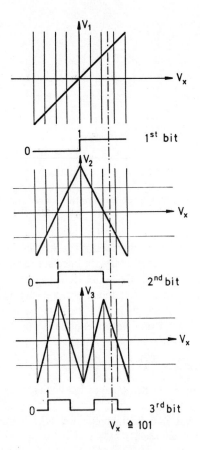

Fig. 3.16 Transfer characteristic for Gray coding

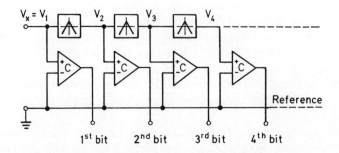

Fig. 3.17 4-bit cascaded Gray code converter

converters. The system is asynchronous, i.e. it needs no clock; hence coding is finished as soon as the last comparison has been made. Since the signal has to propagate through all the cascaded stages in series with markedly different amplitudes of the successive samples, high slew rate performance is a prerequisite. This increases with the number of stages.

There are many ways to implement the V-characteristic in a circuit. Actually, nearly all the known methods use a piecewise composition whereby a non-linear element is introduced to provide the required break-point. In fact, every full-wave rectifier circuit is suitable. However, linearity, precision, and speed are important characteristics that should also be considered. In the following, some circuit realizations are discussed.

Operational Amplifier Realization of a Folding Characteristic

As shown in Fig. 3.18, two operational amplifiers and diodes are used to give the required V-shaped transfer function. The part of the circuit inside the dashed rectangle establishes the characteristic, as shown in Fig. 3.19(a). For $V_k > 0$, diode D_1 conducts and closes the feedback loop containing the resistor R. Due to the unity resistance ratio, the amplification is -1, yielding a slope of $s = -1$ of the characteristic. On the other hand, for $V_k < 0$, the output of the inverting operational amplifier becomes positive and D_1 is reverse biased, keeping the output voltage zero. Accordingly, the feedback loop through D_2 and D_3 is closed. In this case, the transistor T_1 acts as a simple comparator. For $V_k < 0$ it conducts, giving $X_k = 0$ (assuming positive logic). Similarly, if $V_k > 0$, T_1 is off, yielding an output $X_k = V_{cc}$, i.e. high or '1'.

Owing to the resistance ratio of 4:1 in the second operational amplifier circuit, the characteristic of Fig. 3.19(a) is multiplied by -4. The part of the characteristic for $V_k < 0$ remains zero, while that in the range $V_k > 0$ assumes a slope of $s = +4$.

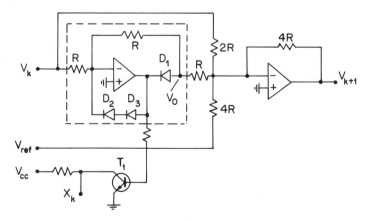

Fig. 3.18   Realization of the folding characteristic using operational amplifiers

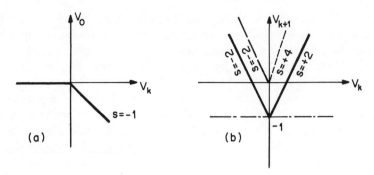

Fig. 3.19 Transfer characteristic of the circuit in Fig. 3.18. (a) Characteristic of dashed block; (b) superposition to composite characteristic

In addition, a linear part having a slope of -2 is added due to the outermost feedback loop that includes the resistor $2R$. The whole characteristic is then shifted downwards by $-V_r$ by a corresponding reference voltage. In this case the input of the second operational amplifier acts as a common summing point.

The resulting overall characteristic, shown in Fig. 3.19(b), could be expressed as

$$V_{k+1} = -2V_k - V_r \qquad \text{for } V_k < 0$$

and

$$V_{k+1} = -2V_k + 4V_k - V_r = 2V_k - V_r \text{ for } V_k > 0$$

It is clear from the above that the output is composed of different signals which flow through different branches. If the delay times and, accordingly, the phase shifts, in these branches are not equal, then the overall characteristic exhibits some distortions, especially at high frequencies. The circuit is therefore suitable only for low-frequency applications.

Another realization that uses a multi-coil transformer without the need for high-precision components is described by Hornak.[25] The measured conversion time for 12-bit resolution is 50 μs.

A Symmetrical Converter for Gray Coding

In order to avoid distortions due to the different propagation path of the signal components, especially at high frequencies, a symmetrical stage-design was introduced by Waldhauer.[26] As shown in Fig. 3.20, a single stage consists of two similar operational amplifier circuits having diodes in their feedback loops in order to provide the knees of the characteristic. A resistance ratio of $2R/R = 2$ is chosen to provide the required slope of the folding characteristic. The circuit is fed by two currents i_P and i_N equal in magnitude and opposite in phase through a differential transconductance amplifier such that $i_P = -i_N = i_1 = g \cdot V_k$.

With positive input voltage V_k, the in-phase current i_P is positive, yielding a negative output of the operational amplifier A_2. This causes D_2' to conduct,

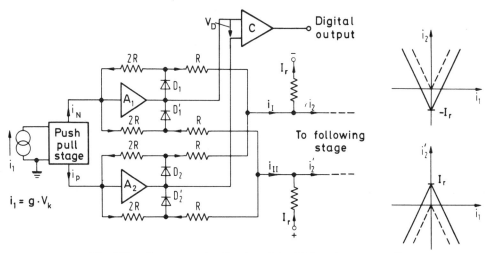

Fig. 3.20 Symmetrical realization of the folding characteristic[26]

producing a voltage drop of $2R \cdot i_P$ across the resistor 2R. A current $2i_P$ is then driven into the resistor R connected between the anode of D_2' and the virtual ground of A_4 (of the next stage). The current i_{II} is given by $i_{II} = -2i_P = -2i_1$. At the same time, i_N is negative and the output of A_1 is positive. By the same argument, diode D_1 conducts and the voltage drop across the resistor 2R drives a corresponding current into R. This leads to $i_I = -2i_N = 2i_1$. Thus in addition to the obtained current amplification of 2 that determines the slope of the characteristic, the two currents i_I and i_{II} are equal in magnitude and opposite in phase.

Similarly, if V_k is negative, the signs of i_P and i_N are interchanged, causing D_2 and D_1' to conduct. This gives rise to the two currents

$$i_I = -(-2i_P) = 2i_1 \text{ and } i_{II} = -(2i_N) = -2i_1$$

It follows, then, that the directions of the two currents i_I and i_{II} remain unchanged, leading to the dotted V-characteristics I and II as shown in Fig. 3.20. The required shift by half the FSR is achieved by injecting the currents $-I_r$ and $+I_r$ to the summing points of the amplifiers A_3 and A_4, respectively. The total currents are thus $i_2 = i_I - I_r$ and $i'_2 = i_{II} + I_r$.

A differential comparator C is connected to the output terminals of A_1 and A_2. Its input voltage V_D, as described above, is $V_D \gtrless 0$ if $i_1 \gtrless 0$. An A/D converter based on the above circuit was built and tested in 1962 for digitizing colour television signals at a 10 MHz sampling rate and a 9-bit resolution. The operational amplifiers were built using vacuum tubes. The trimming of the resistors was tedious, since thin-film resistors with trimming facilities were not known at that time.

A two-stage concept for a 2 MHz word-rate converter with a 15-bit resolution and digital error correction is described by Zimmer.[27] The first stage consists of eight cascaded single folding stages yielding 8-bit. After converting to binary, an 8-bit D/A-converter provides an output for a subtractor circuit to form a residue signal for the next block of eight cascaded folding stages. Wideband operational amplifiers and diode bridges are employed for high precision and folding as well as for subtraction and interstage amplification.

A Folding Characteristic Using Differential Amplifiers[28]

The circuit shown in Fig. 3.21 produces the required folding characteristic with a small number of components and consists mainly of two differential amplifiers. The first is a linear amplifier and contains transistors T_1 and T_2 with a current source and the emitter resistors R_E. T_3 and T_4 actually constitute a current switch rather than an amplifier. The switching at $V_{in} = 0$ is accelerated through the feedback resistors R_F by diminishing the periods during which both T_3 and T_4 are conducting.

With $V_{in} > 0$, the output $V_{2R} - V_{2L}$ is positive. Thus T_3 is off, while T_4 is conducting and acts as an emitter follower. The opposite takes place if V_{in} becomes negative. In this case $V_{2L} - V_{2R} < 0$. Hence T_4 is off and T_3 acts as an emitter follower. The output voltage is taken across R_1 that provides, with R_2, V_{EE}, and the quiescent current through T_3 and T_4, the required shift of the characteristic. The break-point of the function is given by

$$V_{out} = -V_{EE} + \frac{R_1}{R_1 + R_2}(V_{EE} + V_{CC} - I_0/2 R_{C1} - V_{BE3,4})$$

in which V_{BE} represents the emitter–base voltage of transistors T at the quiescent point. The slope of the resulting characteristic is given by

$$dV_{out}/dV_{in} = \pm \frac{R_1}{R_1 + R_2} \cdot \frac{R_{C1}}{2(R_E + r_e)}$$

where

$$r_e = \frac{V_T}{I_0/2}$$

and V_T = volt-equivalent of temperature ($= 26$ mV at room temperature). The negative sign is for negative inputs, and vice versa.

The dashed lines in Fig. 3.22 indicate the shape of the characteristic without feedback by R_F, i.e. a more gradual transition in the vicinity of $V_{in} \approx 0$.

Due to its simplicity and high speed, the circuit seems to be attractive. However, it has the drawback that the slope of the resulting characteristic depends on nearly all the resistors (except R_F and R_{C2}). The same also applies to the shift of the output voltage which depends, in turn, on this slope.

A 6-bit A/D converter using five such stages was built and tested.[28] The sampling pulses to the different stages, as shown in Fig. 3.23, were delayed by a time equivalent to the propagation delay in each stage to implement the flash mode.

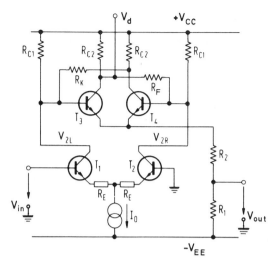

Fig. 3.21 Implementing the folding characteristic with differential amplifiers[28]

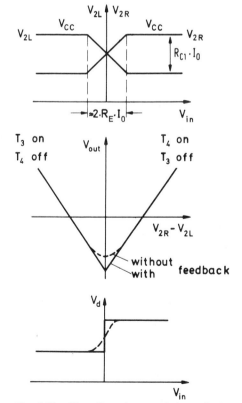

Fig. 3.22 Transfer characteristics of the circuit in Fig. 3.21

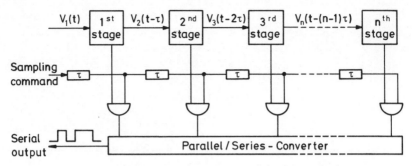

Fig. 3.23 Block diagram of a n-bit converter using the circuit of Fig. 3.22

However, the propagation delay is dependent on the signal frequency. For example, the measured S/N (mainly due to quantization) was reduced by 3 dB at 100 kHz, which corresponds to an error of 1/2 bit. If a sample-and-hold circuit was used instead, the S/N remained constant up to 1 MHz.

Sequential–parallel Gray Code Converters

The realizations described up to now belong to the bit-by-bit conversion type, therefore speed is limited by the slew rate capabilities of the series of stages. Consequently, sequential–parallel structures have also been developed for the Gray code.

Instead of establishing the multiple folding characteristic shown in Fig. 3.18 for three bits by cascading of single folding stages, the multiple folding can also be provided by a parallel circuit configuration. Due to the fact that the number of folded segments to be synthesized increases by a factor of two for each bit, the number of folding segments made in parallel remains limited to 3 or 4 bits. The remaining bits are obtained by a subsequent conventional parallel converter that discriminates a corresponding number of levels along each segment of folded characteristic, or by another multiple folding circuit.

Woods and Zobel[29] use a combination of operational and high-frequency (feedforward) amplifiers in connection with a diode bridge (hard limiter) circuit for each straight-line segment of the folding characteristic. The converter is capable of digitizing a full-range sine wave of 40 kHz for 1/2 LSB degradation at 8-bit resolution and a conversion rate of 30 MHz for tracking applications. 10-bit at 10 MHz and 8-bit at 100 MHz are considered feasible. Fiedler and Seitzer[30] start from the differential amplifier circuit of Fig. 3.21 providing one V-section of a folding characteristic. By putting eight amplifiers in parallel with properly shifted reference voltages and extending the double emitter follower T_3, T_4 to a multiple input emitter follower (voltage selector), multiple folding for 4 bits is achieved. The output is fed to another 4-bit multiple folding unit. By suitable interconnection of the emitter followers' collectors, only eight comparators are needed for the 8-bit converter. Full-range sine wave encoding works up to 1 MHz, and the maximum conversion rate is limited to 14 MHz by the encoding logic.

A similar circuit called 'dual rank flash∗flash' has been recently announced[95] for a digital scope. Two stages (ranks) are cascaded, each of them for 4 bits. The first stage consists of eight folding sections connected in parallel but shifted in reference level. A sampling rate of 100 MHz is quoted and an effective resolution of 4 bits of a 35 MHz full-range sine wave is achieved.

Arbel and Kurz[31] also make use of multiple differential amplifiers. By suitable addition of their output currents a multiple folding characteristic is achieved. The non-linearities of the emitter–base characteristics must be compensated by non-linear scaling of the parallel converter that follows. A converter of 8 bit is reported to work in the tracking mode up to 400 MHz.

Van de Plassche and van de Grift[32] use a combination of bipolar transistors and diodes in a multiple current switch with combination to achieve a 3-bit coarse Gray code quantizer followed by a conventional 4-bit parallel type. The 7-bit converter is fully integrated and works up to a 50 MHz conversion rate.

3.3.5 Pipelining Analog-to-Digital Converters

The input signal to a cascaded digitizer (Fig. 3.10) flows sequentially through the consecutive stages. Since the decision at any stage depends on the result of that preceding it, each stage is engaged only for a fraction of the sampling period. The idea of pipelining converters is simply to produce the n bits during several sampling periods rather than during one period. The result is less stringent speed requirements of the individual blocks but at the expense of an increased conversion time. For many applications, such as video and picturephone, the additional delay is insignificant if compared with the associated delay due to long-distance transmission.

The basic structure of a pipelining converter is illustrated in Fig. 3.24. In view of the fact that during each sampling period the first block receives and encodes a

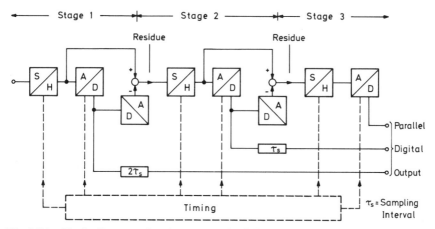

Fig. 3.24 Block diagram of a three-stage pipelining converter using analog delays instead of sample-and-holds

new sample, it is necessary to connect a sample-and-hold element at the input of the second, third, and nth stages. This certainly leads to a loss in the improved speed due to pipelining. In addition, the output bits are not produced simultaneously, so that storage in the form of delay elements or shift registers of different delays for each bit are necessary. Rollenhagen describes a 10-bit converter for a 10 MHz conversion rate consisting of three stages of 5-bit, 4-bit, and 3-bit, respectively, the extra bits being used for error correction. The total delays amount to 300 ns, i.e. three samples of the signal are worked upon at a time. For reasons of economy, analog delays (cables) were used instead of additional sample-and-holds.[14]

Another example of a pipelining converter is a commercially available one, which encodes 10 bits at a 20 MHz rate needing 285 ns conversion time. Additional features are one extra bit for digital error correction, and analog delays between stages of 5-bit and 6-bit, respectively.[33] A related approach, in that time-interleaving is used, is based on an array of four independent 7-bit successive approximation converters each with a sample-and-hold. They work with their sampling commands interleaved in regular time division, and their digital outputs are multiplexed to form a continuous sequence of output words. By this, a number of 4 is gained in time for the individual converter and the die size of the integrated version is reduced as compared with a pure parallel design because $4 \times 7 = 28$ comparators are needed instead of $2^7 = 128$.[34].

3.4 Feedback Analog-to-Digital Converters

3.4.1 Introduction

Feedback or cyclic converters are those in which the signal circulates in a closed loop, i.e. it uses the same blocks in every conversion step. This is in contrast to a cascaded structure, where the signal proceeds from one block to the next. Some authors, however, also use the term 'feedback converters' for cascaded structures, although they are actually provided with feed*forward* rather than feed*back*. Therefore, to avoid confusion, the term 'cyclic' would be more adequate. According to the classification of Best,[1] cyclic converters are those containing pre-subtraction. A subtraction is first made here which is then followed by a comparison from the first estimated value.

The basic structure of a feedback (or servo) converter is shown in Fig. 3.25. The analog input V_x is compared with an already available first estimate $\hat{V}$. The voltage difference V_e, representing the error signal, is then used to produce a new improved estimation. These estimated values are developed in the circuits of the feedback loop. According to the state of the comparator, the block labelled 'logic' controls the counter to start counting in a forward direction for a positive error signal and in a backward direction for a negative error signal. For a modulo-two counter, the output is in binary form. The D/A converter connected in the feedback loop now delivers the required first estimate $\hat{V}$ of the input signal. In the next cycle, this value is then subtracted from the input to give the error signal V_e. The process is terminated if the error signal V_e becomes $|V_e| < Q/2$.

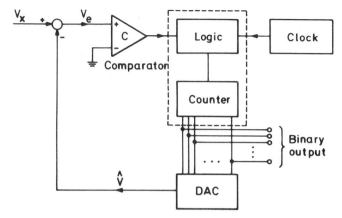

Fig. 3.25 Structure of a feedback converter

However, due to circuit noise and drift, even if $V_e = 0$, a stable state is never maintained, and the counter contents bounces up and down by one LSB in tracking the input value until conversion is stopped.

Since the estimated value successively approaches the input signal until the error is diminished to a predefined value, this method is sometimes also called 'servo technique' when referring to early mechanical motor-driven approaches.

The two blocks labelled 'logic' and 'counter' become relatively simple in the case when the converter is to implement the successive approximation method, i.e. one bit per cycle. Then the two blocks essentially reduce to a successive approximation register (SAR), shown in dashed lines in Fig. 3.25.

There is a distinct advantage to this structure. Any type of D/A converter in the feedback path will allow an A/D converter to be constructed. D/A converters are conceivably simpler than A/D converters. So if the technology easily leads to a D/A converter, the A/D converter is obtained by completing the loop circuit.

The majority of the commercially available A/D converters ($\approx$ 90%) which are used in ordinary digital voltmeters are based on a feedback structure in conjunction with the simple counting method described below. If the cycle time is 1 µs (clock frequency of 1 MHz), then a 3-digit decimal number could be obtained in a few milliseconds, which is satisfactory for many applications.

The realization of this technique in a low-cost form has been favoured for a long time due to the fact that most of the components required are available as standard integrated circuits. A successive approximation A/D converter using standard ICs is illustrated in Fig. 3.26.[35]

As shown in Fig. 3.26, the error signal is produced at the summing node of the currents I_a and I_x, i.e. at the comparator input. The current I_x is obtained from the input voltage V_x via a buffer stage and adjusted with a high-precision trimmer resistor. The current I_a is produced in the D/A converter (described in Chapter 4). The comparator output is given by

$$A = 1 \qquad \text{for } I_a > I_x$$

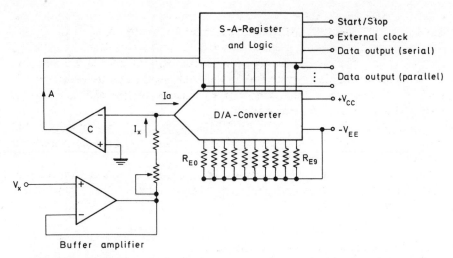

Fig. 3.26 Realization of a feedback converter using standard integrated circuits[36]

and
$$A = 0 \quad \text{if } I_a < I_x$$

In the block labelled SAR, given in more detailed form in Fig. 3.27, these conditions are considered. Thus, upon initiating the clock generator, first the flip-flop No. 9 (FF9) of the register is set to the state $b_9 = 1$ in the first circle through a modulo-11 counter and a decoder; b_9 is then used to drive the current switch of the MSB in the D/A converter. This delivers, in turn, the current I_a for the first

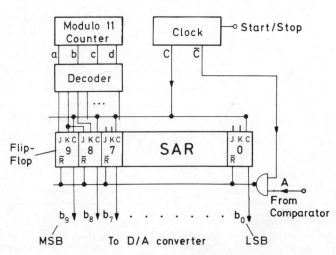

Fig. 3.27 Details of the control logic of Fig. 3.26

comparison. If the state $A = 1$ is obtained at the output of the comparator, then this flip-flop (FF9) is reset to 0 via the reset input R when the clock is low, i.e. $R = 0$. In this case, the current I_a corresponding to 1/2 FSR is removed; otherwise it is left if $A = 0$. In the subsequent steps (cycles), the remaining 9 bits are produced using the same procedures controlled by the clock. At the end, the coded value is obtained at the output of the successive approximation register and the counter is then reset to zero in the eleventh cycle.

A conversion time of 18 μs for the 10 bits is quoted.[35] The maximum and minimum currents of the D/A converters are 1.25 mA and 2.5 μA, respectively. Bipolar ICs are used in the realization.

In the meantime, monolithic integrated MOS-circuits achieve at least comparable results, e.g. 10 bits in 23 μs,[36] as described in section 3.5, but they are less expensive to manufacture.

3.4.2 The Counting Method and its Extension

The analogy to the process of measuring the length X of a stick is considered again for the counting method. The actual length X is expressed as a multiple $x = X/Q$ of one quantizing interval Q. The reference length is assumed equal to the normalized quantum $q_0 = q = 1$. A new variable d is introduced such that, at the beginning $d = q$. The process begins by comparing the normalized lengths x and d. If $d < x$, then d is increased by one quantum to start a new comparison, and so on. The measurement is finished if $d > x$. By counting the number of steps necessary to reach this state, the length x is determined or, by analogy, its equivalent digital representation.

Obviously, only one single reference of size q_0 is used. For a word-length of n bits, the required number of steps, in the word case, is $2^n - 1 = m - 1$. (x will lie in the last quantization interval if it is still greater than d after $m - 1$ steps.) If the reference is inaccurate, then by continously adding q_0 an accummulation of errors takes place that leads to 'gain errors'. In this case, the average slope of the quantization characteristic is not unity.

The counting method is also referred to as the 'level-at-a-time' technique. To overcome the main drawback of the counting method, namely the large number of measuring steps, the so-called extended counting method can be used. In this case, a reference value other than 1 is used. Thus, if the value $q_1 = 2$ is chosen the number of measuring steps is halved at the expense of the resolution. This is taken into account in the last step. As soon as $x < d$, the last increase by $q_1 = 2$ is removed and replaced by $q_0 = 1$ that has to be used only once. Using this improvement, the necessary number of steps for an 8-bit converter, i.e. $m = 256$, is reduced to $127 + 1 = 128$.

Obviously, further improvements can be achieved if the beginning of the comparison is first made with an increasingly larger reference which is then replaced by a smaller one if the state $x < d$ is reached. Thus, if we choose a reference $q_2 = 4$ to start with, then the measurement is finished after a maximum of $k = 63 + 1 + 1$ steps. In general, if N references having the values $1, 2, 4, 8, \ldots, 2^{N-1}$

Table 3.3 Required number of steps depending on the number of references in an extended counting method converter with $m = 256$

No. of references N	Size of references q_i	No. of steps k	Figure of merit $N \cdot k$
1	1	255	255
2	1, 2	128	256
3	1, 2, 4	65	195
4	1, 2, 4, 8	34	136
5	1, 2, 4, 8, 16	19	95
6	1, 2, 4, 8, 16, 32	12	72
7	1, 2, 4, 8, 16, 32, 64	9	63
8	1, 2, 4, 8, 16, 32, 64, 128	8	64

i.e. a binary scaling, then the maximum number of steps is given by

$$k = \frac{m}{2^{N-1}} - 1 + (N-1)$$

If the largest reference has a value that is equal to half the FSR, i.e. $q_{N-1} = 2^{n-1}$, then we get

$$k = \frac{2^n}{2^{n-1}} - 1 + n - 1 = n$$

which is equivalent to the successive approximation, or weighting, method. In Table 3.3 the different possible cases, ranging from the simple counting method to the weighting method, are considered for a resolution of 8 bits.

It is by no means compulsory to scale the references by a factor of 2, as chosen above for the purpose of explanation. There is an analogy here to the extended parallel technique. At a fixed number of references, analogous to a fixed number of stages, a minimum in the number of steps, analogous to the number of comparators, is achieved if each reference is used an equal number of times, analogous to the case that the number of comparators is equal in each stage. For example, take $m = 256$ and $N = 4$ references; in this case their size should be chosen such that they are used the same number of times. This leads to the values $q_0 = 1, q_1 = 4$, $q_2 = 16, q_3 = 64$; each of them is used three times and the total number of steps is $k = 12$. Compare Table 3.2, where for $k = 4$ stages the minimum number of references ($=$ comparators) is $N = 12$. The number of references times the number of steps, $N \cdot k$, introduced as a figure of merit in Table 3.3, gets $N \cdot k = 48$, which is the minimum achievable value.

3.4.3 Circulation 1-Bit per Cycle Analog-to-Digital Converters

A common disadvantage of both the 1-bit per stage cascade (section 3.3.3) and the feedback digitizer with a D/A converter (section 3.4.1) is the fact that during each measuring step only one block of the chain in the first and a certain reference of the D/A converter in the second are used. The converter presented here eliminates the use of a D/A converter as well as the need for several references (actually, two references are required). This is made possible by amplifying the residue signal by a factor of 2 each cycle such that the same reference can be used.

An implementation of this converter is shown in Fig. 3.28. This consists mainly of a single comparator, seven analog switches, an operational amplifier, and n flip-flops. The conversion process is initiated by closing switches S_3 and S_6, leaving all other switches open. The input voltage V_x charges the capacitor C_A via S_3, S_6, and the voltage follower operational amplifier circuit. Due to its high input and low output impedances, the voltage follower does not load the signal source such that the charging of C_A is accelerated. The circuit accordingly has a built-in sample-and-hold circuit. Depending upon the value of V_x, the comparator decides whether it is in the upper or the lower half of the measuring range. Accordingly, a logical 1 or 0 is obtained which is then stored in the flip-flop under the control of the clock. The outputs $\bar{F}$ and F actuate the switches S_1 and S_2 in the following manner:

$$S_1 \text{ is closed if } F = 0 \ (\bar{F} = 1)$$

and

$$S_2 \text{ is closed if } F = 1 \ (\bar{F} = 0)$$

In the first cycle, S_3 and S_6 are open while S_4 and S_7 are closed. If, for example, $F = 0$, i.e. S_1 is closed, then the operational amplifier circuit, whose amplification

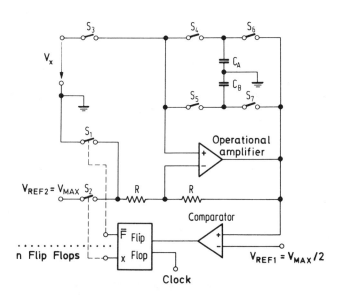

Fig. 3.28 Block diagram of a circulation A/D converter

is selected through S_1, amplifies the voltage V_{CA} of the capacitor C_A, without loading it, by a factor of 2 and charges C_B, achieving a voltage $V_{CB} = 2V_{CA}$. This means that the range is again extended to its full value, making it possible to obtain the next significant bit by using the same reference, i.e. $V_{FS}/2$.

If, on the other hand, $F = 1$, then S_2 is closed and the operational amplifier circuit acts as a difference amplifier, yielding an output of

$$V_{CB} = 2\,V_{CA} - V_{REF2} = 2\,(V_{CA} - V_{REF2}/2) \qquad \text{where } V_{REF2} = V_{MAX} = V_{FS}$$

The comparator output, which is then transferred to the flip-flop upon receiving the clock pulse, represents the second bit. At the same time, the first bit is shifted to the next flip-flop.

In the second cycle, S_5 and S_6 are closed and, according to the state of the flip-flop, S_1 or S_2 are closed. In this case, V_{CB} charges C_A. The first cycle is then repeated.

The advantage of this converter lies mainly in the repeated use of the same small number of components. In addition, no precise capacitors are needed since they are charged to a certain voltage through the operational amplifier. The same applies to the resistors, whose ratio rather than their absolute values determines the transfer function. Finally, the fact that the circuit also implements the function of a sample-and-hold and that the converter's resolution is not, in the first-order consideration, related to its complexity makes this concept attractive.

The main disadvantage is the need for a high-performance operational amplifier. Besides the usual requirements of high open loop gain, small offset voltage, high input impedance, and small temperature drift, its input capacitance is of prime importance. In every cycle the capacitor voltage V_{CA} or V_{CB} is reduced, due to this input capacitance, upon closing S_4 or S_5, respectively. Therefore, capacitors C_A and C_B must be chosen as large as possible. However, this will affect, in turn, the speed of the system due to the increased time constant.[37]

Analog switches must be used here since they must transfer dynamic voltages without distortion. This increases their performance requirements, especially at an increased number of bits. Besides, the required control circuit (not drawn) is not a simple one.

Such a design, which represents, in fact, a compromise between speed and precision, was realized for a 12-bit converter operating at a conversion rate of 8000 conversions per second.[37] Recently, a 16-bit design with a 1.2 ms conversion time based on a circulation of the signal through a folding stage for the Gray code (see Section 3.3.4) has been described.[96]

3.4.4 Examples of Multichip or Monolithic Feedback Converters

High-performance monolithic analog-to-digital converters are difficult to achieve by a single integrated circuit technology. While bipolar transistors are well suited for linear, i.e. analog, circuit functions, compact logic circuits are not feasible due to junction isolation of the devices. Moreover, linear circuits with high breakdown voltages and high current gain also exhibit high capacitance, high saturation resistance, and long storage times when used as logic elements. On the other hand, field effect transistors with their simple structure are ideal for high-density logic and a considerable number of them are needed in an A/D converter for storage, counting, and control. However, their performance as linear circuits is rather poor, and upgrading by using CMOS or junction FETs in addition means a more complex technology and/or less efficiency in silicon area usage.

It is interesting to see what answers are given by the technology at the beginning of the 1980s. There are a number of them, as we will see, ranging from limited performance over multiple-chip approaches and mixed bipolar/FET technology to trimming and/or error correction and self-compensation techniques leading to signal processing on the chip or special coding algorithms, as will be discussed in subsequent chapters.

A monolithic 10-bit A/D converter based on the successive approximation feedback structure has been made[38] based on the bipolar collector diffusion isolation (CDI) technology. This consists of a 10-bit D/A converter, a successive approximation register, and a reference and reference amplifier as well as control logic on a chip of about 4 mm × 4 mm. The conversion time is 10 µs. The CDI process yields high packing density and includes analogue and digital circuitry. Another approach, based on the same technology, is aimed as a tracking converter,[39] i.e. it implements the straightforward counting method. This is a 10-bit unit, working at a 1.5 MHz clock rate, which is said to give a 10 : 1 improvement in bandwidth over successive approximations for certain small-signal applications, often obviating the need for a sample-and-hold circuit.

Integrated injection logic (I^2L) is another contender for high-density bipolar circuits. This is used for a 10-bit monolithic successive approximation feedback converter with on-chip comparator, clock, and buried Zener reference.[40] Accuracy is high due to laser wafer trimming, and its conversion time is 25 µs. With few additional components, the unit operates from standard microprocessor control lines, presenting its data to standard bus formats between a 4- and 16-bit width. This design has been up-graded to allow for a 12-bit/25 µs converter on two chips within a 28-pin DIP, representing a new landmark for monolithic realization of data conversion building blocks.[99]

Another I^2L feedback successive approximation chip converts 13-bit in a time of 16 µs.[41] It includes a sample-and-hold circuit with a PMOS transistor as a switch. Ion implantation is used to reduce the PMOS threshold voltage.

High speed at low cost for video applications is the objective of preliminary work on a feedback successive approximation converter aimed at a 15 MHz sampling rate for 8-bit and 6 MHz for 10-bit resolution.[18] The key to high speed is a fast DAC using switched current sources (see Chapter 4) of bipolar technology with settling times of 5 ns/12 ns for 8-bit/10-bit, respectively, and a successive approximation register capable of working up to a 135 MHz clock rate.

An integrated NMOS A/D converter of the feedback successive type has been shown for 8-bit resolution and a conversion time of 13.2 µs. For the D/A converter, a current source array is used to achieve high speed and the problems associated with trimming resistors for the weighting of current sources (see Chapter 4) are avoided by using MOS transistors.[42]

Recently, the high-resolution field has been tackled by using integrated circuits. Three chips (including CMOS, complementary vertical bipolar transistors, and laser trimmed nichrome resistors) for the one digital circuit chip and complementary vertical bipolar and implanted junction FETs, trimmed nichrome resistors, and dielectric isolation for the two analog processor chips are used for making a 12-bit A/D converter for a 30 µs conversion time.[92] The algorithm is of the successive approximation type, and processing between the two analog processing chips is similar to that in Fig. 3.28 between the capacitors C_A and C_B. The three chips are assembled within one 40-pin DIP.

Another approach to 14-bits resolution on one chip at a conversion time of 20 µs is based on CMOS technology. Instead of trimming, an error-correcting algorithm in connection with a redundant D/A converter and a successive approximation register in the feedback path is used.[43] Seventeen bits of internal resolution are needed to compensate for accumulated errors. The result is rounded off to 14 bits. Errors in the high-order bits are corrected by subsequent addition of lower significant bits. This is possible due to a weighting of 1.85 instead of 2, yielding an overlap for correction.

For data acquisition and conversion, a monolithic 12-bit device for a 3 µs conversion time has been developed based on a 1 Ωcm, 3.8 µm epitaxial process yielding 1.5 GHz NPN transistors, Schottky diodes, and polysilicon resistors. The linear differential logic (LDL) has typical gate delays of 3 ns, and the logic delay per bit is 30 ns. The comparator responds to $Q/2$ overdrive in 10 ns. Therefore the major limitation for conversion speed is voltage settling time at the summing node. The D/A converter obtains its linearity through a combination of fifteeen weighted currents for the four most significant bits and a R–2R resistor ladder and active emitter scaling for the remaining 8 bits (see sections 4.2 and 4.3). A voltage reference is included on chip by means of a corrected bandgap reference (see section 5.4.2).[97]

3.4.5 Interpolation Analog-to-Digital Converters

The motive for applying interpolation techniques to data converters, as suggested by Candy,[44] is to achieve economy by trading speed for resolution. Savings are obtained by a smaller number of components. Of course, this works only up to a certain limit beyond which a further increase in speed costs more than the savings on components, i.e. up to about a 30 MHz clock frequency.

A block diagram of an interpolation A/D converter is shown in Fig. 3.29. This consists mainly of two parts: a feedback part for coarse quantization at a quantum size of Q and an interpolator to refine the coarse quantizing of the first part by a factor N. For this purpose, the sampling period is divided into N clock periods of duration τ. The read-out of the digital accumulator, taken each N clock-period (i.e. once per sampling period), represents the digital output.

The operation of this converter is best illustrated by considering the example given in Table 3.4. Assuming that the integrator output increases by one quantum

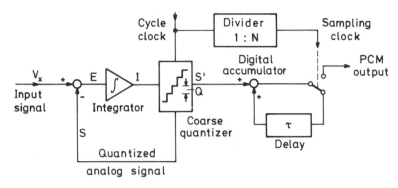

Fig. 3.29 Configuration of an interpolative A/D converter

Q within the time τ if its input, i.e. the error E, amounts to $E = Q$, and normalizing all values to Q, then for an analog input of 8.2 and an initial integrator output $I = 5.1$ the quantizer output is $S = 5$. (In Fig. 3.29, S' is the digital representation of S.) The error signal fed back to the integrator input is now $E = 3.2$. After one time interval τ, this gives rise to an input to the quantizer of $I = 8.3$ and, accordingly, an output of $S = 8$ (the threshold level lies upon 8.5 between each two successive levels). The process continues as indicated in Table 3.4, such that, after five periods, the first value of the integrator output, i.e. 8.3, is reached again. The sum of the quantizer output sequence of 8, 9, 8, 8, 8 divided by $N = 5$ delivers the value 8.2. Changing the starting value leads only to a shift in this sequence, but their sum and the proportion of occurrence of the upper value 9 remain unchanged.

The interpolation part contains a divider $(1/N)$ which is adjusted such that the output of the digital accumulator is taken every N cycles. If the divider was the modulo-two type, then the division by factor N could be carried out simply by shifting the output by $\log_2 N$ digits.

Based on this method, an 8-bit A/D converter was realized for a picturephone system with sixteen quantization levels. The signal bandwidth is limited to 1 MHz such that a sampling rate of 2 MHz is adequate. With $N = 16$ the necessary clock

Table 3.4 Example of interpolative quantization ($Q = 1$, $N = 5$, $1 \leq n \leq 5$)

Analog input V_x	Error $E_i = V_i - S_{i-1}$	Integrator output $I_i = I_{i-1} + E_i$	Quantizer S_i	Output $A = \dfrac{1}{N} \cdot \Sigma S_i$
	Starting →	5.1	5	
8.2	3.2	8.3	8	Sampling interval
8.2	0.2	8.5	9	$T_s = 5 \cdot \tau$
8.2	−0.8	7.7	8	
8.2	0.2	7.9	8	$(8 + 9 + 8 + 8 + 8) : 5 = 8.2$
8.2	0.2	8.1	8	
8.2	0.2	8.3		

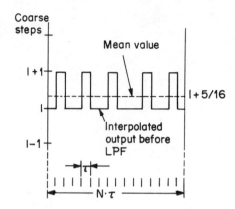

Fig. 3.30 Interpolative D/A converter signal for $V_{out} = l + 5/16$ before averaging

frequency is 32 MHz. In a modified version, non-linear quantization using the μ-law (section 3.6.3) with $\mu = 255$ was achieved for a constant signal-to-noise ratio over a wide range of amplitudes.[45]

The same technique can be applied to design D/A converters.[46] An ordinary D/A converter produces coarse levels $l - 1, l, l + 1$ (Fig. 3.30). According to the proportion of occurrence of the higher level $l + 1$, a certain number of pulses having the amplitude of a coarse interval q are then added to the value l. In Fig. 3.30, the proportion is 5/16. A low-pass filter connected to the output is used as an analog interpolator to deliver the average amplitude value.

The filter time-constant could be determined by using the corresponding values of a stochastic D/A converter (section 3.6.5), i.e. by using $t_p = N \cdot \tau$. This fact indicates a relationship between this type and the stochastic conversion technique. In the latter, interpolation is made by the statistical occurrence of pulses describing the signal by their mean value. Because the statistical approach treats each measurement as an independent event without using the results of previous measurements (as the interpolative method does by integration and coarse quantization) many more events are needed in the statistical case.

The process of interpolative conversion can also be considered in the frequency domain. Due to the oversampling by a factor N, the spectrum of the quantizing noise power is distributed over a wider frequency range. Due to the integrator, the spectrum of the quantizing error is shaped to increase at higher frequencies, and as a result of the interpolator at the output operating as a low-pass filter, the quantizing noise power falling into the frequency band B of the signal is reduced by a factor $N/\sqrt{2}$.[47]

The spectral shaping elements, i.e. the integrator and the interpolator, can be altered to improve further the signal-to-noise ratio, i.e. to simplify further the quantizer section.[48] In this case the interpolator should be replaced by an ideal low-pass filter with a bandwidth B of the signal to be encoded. This is interesting, because then the low pass in front of the A/D converter can be much simpler. It has only to remove aliasing components, which is easily done in the case of considerable oversampling. The integrator should be of a higher order. For reasons of stability, this should be restricted to second order.

Based on the considerations above, a one-step quantizer is able to encode a telephone signal with a bandwidth $B = 4\,\text{kHz}$ by oversampling to $512\,\text{kHz}$, corresponding to $N = 64$, at the same signal-to-noise ratio as 12-bits linear PCM.[47] A similar approach has been proposed in that the signal spectrum is shaped before being quantized.[49] In addition to oversampling and noise shaping, linear prediction[50] and non-linear quantization can be used to save in coder complexity or the number of bits for encoding (see sections 2.2 and 3.6).

3.5 Indirect Converters

All the conversion techniques discussed in the previous sections perform the measuring process directly by comparing the input signal amplitude with one or more references, either voltage or current. In indirect converters, on the other hand, the input voltage is first transformed into another auxiliary quantity which, in the sense of the encoder to be built, can be measured in a way which is cheaper, more accurate, or faster. Examples of such auxiliary quantities are frequency and time. Also, due to the recent development in MOS-integrated circuit technology, where high-quality precise capacitors are produced, charge rather than time or frequency is being considered for carrying the analog information during the coding process. It is possible, therefore, to convert the input signal level to, for example, a proportional frequency. The number of periods of this frequency, counted over a predetermined interval, represents the signal amplitude. Alternatively, the analog level could be converted to a proportional time period over which the number of cycles of a fixed frequency are counted to provide the result. Basically, this is another implementation of the counting method (level-at-a-time), as described in section 3.4.2. By suitable modifications, as we shall see, the extended counting method is implemented.

3.5.1 The Single-Ramp Method

The basic structure and the associated voltage waveforms for a circuit implementation of this method are shown in Fig. 3.31. The input voltage V_x is compared with a linear voltage function (ramp) V_r in the comparator C1. The output V_{C1} remains high as long as $V_x > V_r$. Comparator C2, on the other hand, compares V_r with ground potential. Hence its output V_{C2} is high all over the ramp period, i.e. for $V_r > 0$. The voltage V_G that results from AND-ing the two outputs V_{C1} and V_{C2} is then applied to a further AND-gate together with a clock. A counter receives a number of pulses only over the period in which $0 < V_r < V_x$. If a constant slope of V_r is assumed, then the counter reading is the binary coded value of V_x.

A ramp function can be generated, for example, by the circuit of Fig. 3.32.

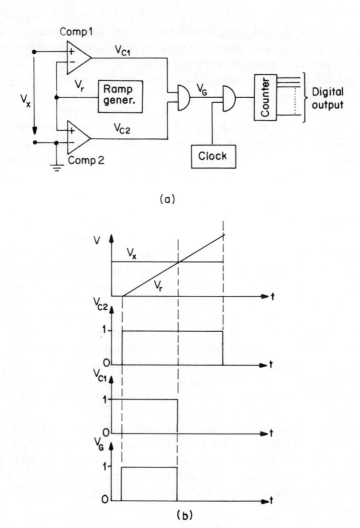

Fig. 3.31 Single-ramp conversion. (a) Block diagram; (b) voltage waveforms

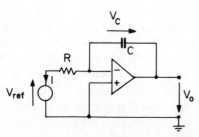

Fig. 3.32 Basic circuit for generating a ramp-function

Capacitor C is charged here by an almost constant current, produced through the constant voltage V_{ref} and the resistor R. Due to the virtual ground at the operation amplifier input, the charging current is constant and the output voltage V_0 is actually increasing linear with time:

$$V_0 = \frac{V_{ref}}{R} \cdot \frac{t}{C} \quad (V_0 = 0 \text{ at } t = 0).$$

At $V_0 = V_x$, then $t = t_M$, where

$$t_M = \frac{V_x}{V_{ref}} \cdot RC = \frac{R \cdot C}{V_{ref}} \cdot V_x$$

i.e. time t_M is proportional to V_x.

It is to be noted that the precision of the components R and C as well as the linearity of the ramp voltage determine the precision of the converter. In addition, instantaneous noise voltage superimposed on V_x may affect the results. Since the counter reading is given by $z = t_M \cdot f_{clock}$, the stability of the clock frequency is also important. For high resolution, the clock frequency must be sufficiently high. Due to these reasons the precision and resolution of the single ramp method is limited to about 10 bits, i.e. a 0.1% error. The minimum time required for one conversion is given by $2^n/f_{clock}$. The reset time of V_r has to be added to this value when calculating the conversion rate.

3.5.2 Dual-Ramp Converters

The operation of the dual ramp or dual slope converter is illustrated by considering the basic block diagram and the voltage $V_0(t)$ at the output of the operational amplifier given in Fig. 3.33.

During the constant time interval t_1, which is independent of the input voltage $V_x(t)$, the capacitor C is charged at a constant rate from the input signal $V_x(t)$ via the resistor R. Thus, the corresponding output voltage is given by

$$V_0 = \frac{Q_0}{C} = \frac{1}{RC} \int_0^{t_1} V_x(t) \, dt$$

For V_x = constant, this reduces to $V_0 = (V_x/RC)t_1$. Instantaneous noise voltages superimposed on V_x are integrated so that their influence is eliminated. Also, hum voltages are cancelled if t_1 is chosen to be an integer multiple of the period of the mains frequency. At $t = t_1$, the block labelled 'logic' turns the switch to the reference voltage V_{ref} instead of V_x. At the same time it enables the counter Z_2, through the AND-gate, for counting. Capacitor C is then discharged in the interval t_2 such that for $V_0 = 0$

$$Q_0 = V_0 C = \frac{V_{ref}}{R} \cdot t_2$$

The counter Z_2 is disabled when a logic '0' is detected at the output, i.e. when $V_0 = 0$.

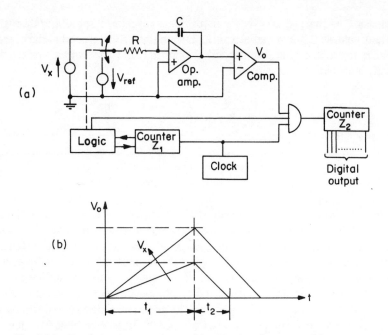

Fig. 3.33 Dual-slope converter. (a) Block diagram; (b) voltage waveforms

Combining the equations above, we get

$$t_2 = \frac{V_0}{V_{ref}} \cdot RC = \frac{V_x}{V_{ref}} \cdot t_1$$

The result, now, is independent of both R and C. The counter reading is given by

$$z_2 = t_2 \cdot f_{clock} = \frac{V_x}{V_{ref}} \cdot t_1 \cdot f_{clock}$$

where t_1 is given, in turn, by the reading z_1 of another counter Z_1 determining the time t_1 as follows:

$$t_1 = z_1 / f_{clock}$$

which leads to

$$z_2 = z_1 \cdot V_x / V_{ref}$$

The clock frequency in this case needs to be constant only during the time interval $t_1 + t_2$. Then it eliminates itself in the counter reading z_2. A resolution up to 14 bits, i.e. an error of 0.01%, is obtainable with this technique.

A major drawback of this method is its relatively large conversion time, especially at high resolution. An improvement in speed is attainable if the so-called 'triple-ramp' method is used.[94] Two discharging currents, rather than one, are provided to be applied during the ramp-down period t_2. This results in two different integration rates and implies that the counter should comprise two counting weights each for a corresponding part of the output word. During the first phase, fast discharging is used when applying the clock to the most significant bits of the counter. Discharging is switched to the low value as soon as the output voltage V_0 decreases to a value below one unit of the higher-order bits. Then the clock is connected to the lower significant bits of the counter.

This converter architecture retains the advantages of the dual-slope one (e.g. inherent linearity without precision components) while increasing significantly the conversion speed at the expense of some added complexity. Obviously, this is an implementation of the extended counting method using two references.

3.5.3 Voltage-to-Frequency Converters

Voltage-to-frequency (V/f) converters, as mentioned in section 3.5, are similar to voltage-to-time or ramp converters: if an analog quantity, such as voltage, is converted to a proportional frequency and the periods of this frequency are counted over a fixed time, the counter output corresponds to the analog value. V/f converters, therefore, share many properties of ramp encoders such as monotonicity, linearity, differential linearity, and noise suppression due to integration. Of course, they are also relatively slow. The key to superior performance are constant currents (see section 5.5) that charge capacitors between stable reference voltages (see section 5.4).

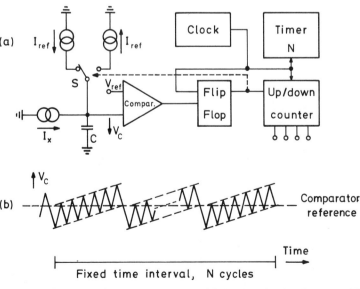

Fig. 3.34 Charge balancing converter. (a) Basic circuit diagram; (b) voltage across capacitor versus time

As a salient representation of the V/f conversion the charge-balancing converter is described. The charge-balancing technique is also known as 'quantized feedback converter' or 'sigma-delta-modulator'. Similar to the dual slope converter, this uses the principles of integration and charge equalization.

The dual slope converter shown in Fig. 3.33 has a two-phase operating cycle, i.e. a measure interval and a count interval. The quantized feedback converter performs the measuring and counting functions at the same time. Figure 3.34(a) shows the basic circuit, Fig. 3.34(b) the voltage across the capacitor C that develops when charging and discharging it over a fixed time interval of N clock cycles. The charging occurs under the control of the switch S by means of the current $(I_x + I_{ref})$, the discharging occurs due to the current $(I_x - I_{ref})$. The switch S is operated by the comparator output and the flip-flop in such a way that the voltage across the capacitor is kept around the comparator reference voltage and the time is quantized in units of clock cycles providing the quantized feedback operation. An up/down counter stores the difference $N_{up} - N_{down}$ of the numbers of cycles that the switch was in the charge or discharge position. This difference is proportional to the input signal I_x.

At small values $I_x \ll I_{ref}$, the up and down slopes of the individual sawtooth remain almost unchanged; and only their centre line slope deviates slightly from zero, leading to additional balancing steps. In the picture, such a step occurs at about every seventh up/down pattern, indicating an input signal close to 1/7 of full scale. If I_x becomes larger, the difference of the slopes within the sawtooth pattern according to Fig. 3.34(b) becomes visible, leading to $N_{up} - N_{down}$ differences within the individual up/down cycle for coarse balancing, the fine balance still being due to additional steps between several groups of up/down cycles. The charge Q_{up} during the charging cycles $N_{up} \cdot t_{clock}$ is

$$Q_{up} = N_{up} \cdot (I_x + I_{ref}) \cdot t_{clock}$$

In the same way, the discharge takes place for $N_{down} \cdot t_{clock}$ cycles, yielding a discharge

$$Q_{down} = N_{down} \cdot (I_{ref} - I_x) \cdot t_{clock}$$

For net charge of zero during the measuring cycle $(N_{up} + N_{down}) \cdot t_{clock}$, $Q_{up} = Q_{down}$, resulting in

$$I_x/I_{ref} = (N_{down} - N_{up})/(N_{down} + N_{up}) = (N_{down} - N_{up})/N$$

The resolution is determined by the length N of the measuring interval.

For comparison with the dual slope principle, an idea of the total voltage for charge and discharge is obtained when adding up linearly all charge and discharge cycles, as shown in Fig. 3.35 for the cases $I_x > 0$ and $I_x = 0$. For $I_x = 0$, the slopes of charge and discharge are equal; for $I_x > 0$, the slope increases by I_x/I_{ref} for charge and decreases by I_x/I_{ref} for discharge, resulting in a difference $N_{down} - N_{up}$ proportional to I_x that can be concluded by geometrical relationships. In a converter these voltage excursions are folded many times around the level of the reference voltage.

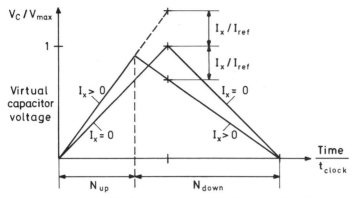

Fig. 3.35 Virtual voltage across charging capacitor C

Comparing dual slope and charge balancing, there are a number of advantages for the latter. Due to simultaneous integration, short-term drift in the clock frequency does not always affect measurement. The dynamic range of the integrator can be very small, and is independent of the value measured. For the same reason, dielectric absorption produces no error. The conversion time interval is fixed and the input signal can be bipolar. It is easy to include offset cancellation by adding auto-zero intervals. Of course, both dual slope and charge balancing can have excellent normal mode rejection by adjusting clock frequency to become a multiple of the main frequency.[51]

The charge balancing converter has been implemented as a two-chip solution, one of them carrying the analog section, the other the digital control circuitry. In one case, the analog chip was based on bipolar transistors, the digital one on LOC-MOS technology.[52] A maximum resolution of 5 digits plus sign (f_{clock} = 200 kHz) and a linearity error of 10^{-5} of full scale (+1 V) was achieved consuming 27 mW from a +7.5 V power supply. Chip sizes were 1.8 × 2.9 mm², and 2.3 × 2 mm². Another implementation took a low-voltage CMOS process for the analog chip.[53] This has 3-digit accuracy and a 20 kHz clock frequency. With a resolution of 1 in 2000, linearity is well within ±1/2 LSB, and it delivers 50 samples per second from a +1 V input range. Power dissipation from +5 V, −5 V supplies are 30 mW and 20 mW and chip size is 3.3 mm².

3.5.4 Charge Redistribution Converters

As an alternative to time and frequency, the charge can also be used as an auxiliary quantity in indirect A/D converters. Intermediate results are stored dynamically with minimum losses on high-precision capacitors (e.g. MOS capacitors), and are moved from one capacitor to another by MOSFET switches. In fact, this represents a step towards economical realizations, since MOS-circuits are now widely produced using monolithic integrated-circuit techniques. Several

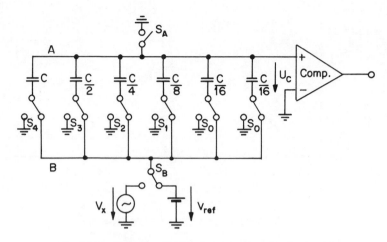

Fig. 3.36 Charge redistribution converter (basic circuit)

approaches to charge redistribution converters were proposed, and one of them, based on successive approximation, is presented here.[36] A different design that is directly suitable for a D/A converter, as described in section 4.5, can be introduced into a feedback structure to provide a cyclic A/D converter.[54]

A simplified circuit of a 5-bit charge redistribution converter is illustrated in Fig. 3.36. The capacitors have binary weighted values, i.e. $C, C/2, \ldots, C/2^{n-1}$. As shown in the figure, two capacitors having the value $C/2^{n-1}$ are connected so that the total capacitance of the $n + 1$ capacitors is $2C$. MOS-transistors are used to implement the required switches ($n + 3$ switches). A voltage comparator provides appropriate steering of the switches via auxiliary logic circuitry.

The conversion process is performed in three steps. In the 'sampling mode', switch S_A is closed and S_B is switched to the input voltage V_x. The remaining switches are turned to the common bus B. Due to charging, a total charge of $Q_x = -2C \cdot V_x$ is stored on the upper plates of the capacitors.

In the 'hold mode', switch S_A is opened while the switches $S_4, S_3, \ldots, S_0, S'_0$ are connected to ground. In the case of no leakage, the charge on the capacitors remains constant and a voltage of $V_C = -V_x$ is accordingly applied to the comparator's input. This means that the circuit already has a built-in sample-and-hold element. Finally, in the 'redistribution mode', where the actual conversion is made, capacitor C (the largest one) is connected to the reference voltage V_{ref} corresponding to the FSR via S_4 in order to establish a 1:1 capacitance voltage divider. The comparator input voltage becomes $V_C = -V_x + V_{ref}/2$. Thus if $V_x < V_{ref}/2$, then $V_C < 0$ and, accordingly, the MSB $b_4 = 0$. On the other hand, if $V_x > V_{ref}/2$, then $V_C > 0$ and then $b_4 = 1$. In the first case (i.e. MSB is low), switch S_4 is turned to ground again to discharge C; otherwise it remains connected to V_{ref} if b_4 is high. In the next step, capacitor $C/2$ is switched to V_{ref}, giving rise to a comparator voltage of

$$V_C = -V_x + b_4 \cdot V_{ref}/2 + V_{ref}/4$$

According to this voltage, the next significant bit b_3 is obtained by comparing V_x to $1/4 V_{ref}$ or $3/4 V_{ref}$ through the voltage divider. Switch S_3 is then either turned to ground $b_3 = 0$, thereby discharging $C/2$, or remains connected to V_{ref} ($b_3 = 1$). The process continues until all bits are generated.

It is to be noted that the circuit is insensitive to parasitic capacitances introduced by the switches. A 10-bit monolithic integrated charge redistribution converter on a chip area of 5 mm² demonstrates its feasibility. The required conversion time for the 10 bits is 23 μs.[36] Further reduction in chip area is possible through savings on the number of capacitors used, as will be shown in section 4.5. More recently, a 12-bit version with a conversion time of 50 μs has been described,[55] as well as an application for a PCM Codec.[56] Proposals have been made to reduce the number of capacitors by ladder network voltage dividers.[100]

3.6 Miscellaneous Analog-to-Digital Converters

Converters presented here differ more or less from the more straightforward ones described earlier by their structure, conversion algorithm, output code, operation, and function. For instance, the weighing method (the put and take technique) can be implemented in a cascaded and/or a cyclic structure. Various structures can also be used for time-multiplexed converters.

3.6.1 The Put and Take Technique

As mentioned earlier, the realization of this technique is not related to a certain structure. Thus the cascaded structure due to Smith[23] realizes the method as equally well as the cyclic (feedback) converters based on either static[35] or dynamic intermediate storage.[36,37] Because only a single bit is generated per step, this method is also sometimes called the 'bit-at-a-time' technique.

More than one reference, having values that differ by a factor of 2 (e.g. $q_0 = q_1/2$, $q_1 = q_2/2$, etc.), is required here. The largest reference q_{n-1} therefore has the value of half the full-scale range.

The conversion process starts by 'putting' the largest reference q_{n-1} to be compared with the input voltage V_x or $v_x = V_x/Q$, respectively. If $v_x > q_{n-1}$, the MSB b_{n-1} is '1'; otherwise it is '0' and q_{n-1} is removed ('taken'). In the second step, put q_{n-2} as next reference and add it to $b_{n-1} \cdot q_{n-1}$. This means that if

$$v_x > b_{n-1} \cdot q_{n-1} + 1 \cdot q_{n-2}, \text{ then } b_{n-2} = 1$$

or

$$v_x < b_{n-1} \cdot q_{n-1} + 1 \cdot q_{n-2}, \text{ then } b_{n-2} = 0$$

As in the first step, if $b_{n-2} = 0$, then the corresponding reference is taken away; otherwise it is left. The process terminates after all references have been employed once. The required number of steps i in this case equals the number of references, i.e. $i = n$. The same resolution as in the simple counting method is obtained in the last step, since in this case the smallest reference $q_0 = 1$ is used. The converter covers the range $m = 2^n$ *with n* references. The required number of steps is given by $i = n = \log_2 m$.

The terms 'put and take' or 'weighing method' stem from the fact that its operation is analogous to the weighing process using a mechanical balance. The references here are standard weights having binary or decimally scaled values. The process starts by placing the largest weight on one side of the balance. If the balance remains in favour of object to be weighed, then this weight is left on and a smaller one is added. If the balance turns in favour of the weight, the weight is removed (taken) and a smaller one is tried (put). Proceeding in this way down to the smallest weight until a balance is reached, the sum of the remaining weights on the pan will represent a closest approximation of the weight of the unknown object.

3.6.2 Non-linear Analog-to-Digital Converters

In all the preceding techniques the value q of the normalized quantization interval (quantum) was assumed constant and independent of the absolute value of the signal. The resulting absolute error is correspondingly constant, whereas the relative error depends on the input amplitude and increases with decreasing amplitudes. The instantaneous signal-to-noise ratio, S/N, varies. However, in some applications it is desirable to keep it almost constant over a wide range of input amplitudes ('robust' quantization). One way of doing this is to vary the quantum size in proportion with the signal amplitude, i.e. to introduce an exponential relationship.

A direct example of these applications is the digital transmission of speech by means of pulse code modulation (PCM). A constant S/N ratio over the whole range is of importance, since it ensures that all speakers, independent of their sound level, are handled with almost equal fidelity of signal reconstruction, due to the fact that the masking threshold of the ear increases with increased sound level such that a louder signal can mask a proportionally louder noise. In addition, the use of non-uniform quantization in PCM telephone links is useful due to more efficient use of channel capacity. As will be shown, a word length of only 8-bit with non-linear quantization yields the same transmission quality as 12- or 13-bit with linear quantization for small signals.

A first realization of such a technique was put into service by the ATT-company in 1962 for PCM links under the name 'T1-system'. As shown in Fig. 3.37, a linear or uniform quantizer is preceded by what is called a 'compressor'. This memoryless non-linearity has the function of shrinking large amplitudes of the voltage V_1 into a smaller range to spread low amplitudes over a wider region, thus allocating more quantizer levels to the lower amplitudes. This has the characteristic given in Fig. 3.37(b). At the receiving end, the signal, having been D/A converted, is applied to the inverse non-linearity within an 'expandor'. The required characteristic, shown in Fig. 3.37(c), is an exact complement (inverse) to

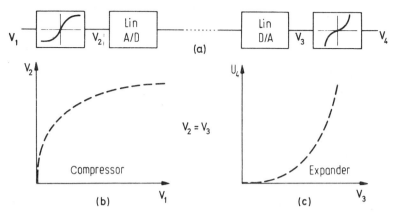

Fig. 3.37 A converter block diagram (a) for non-uniform quantization including companding characteristics (b), (c)

that of the compressor. The transmission, in this case, is effectively linear. The process of compressing and expanding is abbreviated as 'companding'.

As mentioned above, an exponential or, depending on the input variable, logarithmic relationship is suitable as a companding law. However, since the function $y = \ln x$ does not include $x = 0$, $y = 0$, some sort of approximation must be introduced.

In an approach suggested by Bell Laboratories (the μ-law), the function takes the form

$$y = \frac{\ln(1 + \mu x)}{\ln(1 + \mu)},$$

in which x and y are the normalized input and output voltages, respectively, i.e.

$$x = V_1/V_{1max}, y = V_2/V_{2max}$$

This function has the following properties:

$$y(x = 0) = 0, y(x = 1) = 1, y'(x = 0) = \mu/\ln(1 + \mu)$$

In Fig. 3.38, the function is drawn for $\mu = 100$ and compared with the function $y = x$ representing linear or uniform quantization. Increasing the slope of a linear quantization characteristic by a factor of 2 is associated with doubling the available quantizing levels, resulting in an improvement of the S/N ratio by a factor of 2, or 6 dB. In the case of the μ-law with $\mu = 100$ and the slope $y'(x = 0) = 21, 7$, the improvement in the S/N ratio for small signals (also called 'companding advantage') is 4.62 · 6 dB = 26.7 dB, which is 4-bit. In practice, the constant μ for more companding advantage is limited. This is due to the fact that the transfer function of both compressor and expandor in the T1 system were piecewise approximated through segments of diode characteristics. Increasing μ means increased curvature of the characteristic, so realization becomes more difficult. This is indicated by the fact that the temperature of both the compressor and expandor in a T1 system has to be kept constant within 0.1°C.

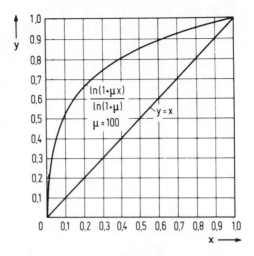

Fig. 3.38 Companding according to the μ-law

In this system, although only a 7-bit word length is used, the performance obtained for small signals is equivalent to that of an 11-bit linear system.

In the more recent version of the T2 system, a μ-value of 255 is made possible by use of digital companding techniques.[45] In this case, $y'(x = 0) = 47$ and the companding advantage amounts to 33 dB.

In Europe, however, another approach has been used to overcome difficulties of the complementary compressor–expandor characteristics for digital switching exchanges on an international scale. The European Committee of Post and Telegraph Administrations (CEPT) has accommodated a robust logarithmic characteristic, the A-law, which is composed of two parts:

$$y_1 = \frac{A \cdot x}{1 + \ln A} \qquad \text{for } x < \frac{1}{A}$$

and

$$y_2 = \frac{1 + \ln Ax}{1 + \ln A} \qquad \text{for larger values}$$

Taking $A = 87.6$, the slope in the origin $y'(x = 0) = 16$ and a companding advantage of $(\log_2 16) \times 6 \text{ dB} = 24 \text{ dB}$ is obtained. This function is piecewise approximated using thirteen segments that are symmetrical relative to the origin. Six segments cover the range $x > 0$, and an equal number for the range $x < 0$. The thirteenth segment, which passes through the origin, extends for input levels of 2 ($x = 1/64$) up to 32 intervals at the output ($y = 2/8$), i.e. it has double the height of the other segments. Therefore the chord numbers 1 and 2 are assigned to this segment (Fig. 3.39). As shown in Fig. 3.39, the slope s of the consecutive segments is successively halved. The thirteen segments are coded by 4 bits, i.e. one for the sign plus three for the respective segment. It is further recommended by the CEPT to subdivide each segment by sixteen equal intervals, taking additional 4-bit. Thus with 8-bit in total, the same quality for small amplitudes is obtained as if 12-bit with linear quantization were used.

Taking the μ-law for $\mu = 255$, the piecewise linear approximation obtained fifteen segments due to subdividing that segment crossing the zero of the A-law into two pieces, the lower one getting a slope of 32, the adjacent one of 16 again.

The signal-to-noise ratio S/N, as shown in Fig. 3.40 for a sine wave signal as a reference, for the 8-bit code format can be constructed starting from 8-bit linear format that has a maximum of $8 \times 6\,dB + 1.8\,dB$, which is valid for the segment having a slope of 1. This segment extends between $-18\,dB < V/V_{max} < -12\,dB$. For

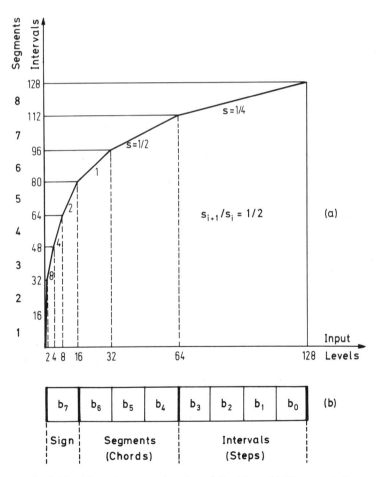

Fig. 3.39 Piecewise approximation of the A-law. (a) Characteristic; (b) 8-bit code format

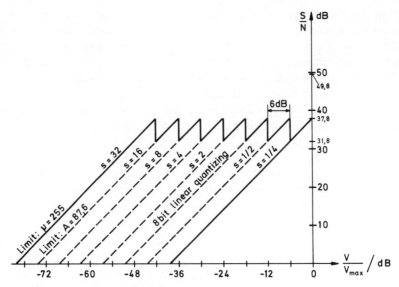

Fig. 3.40 Signal-to-noise ratio for A-law and μ-law companding characteristic (sine wave signal, no overload)

$V/V_{max} > -12$ dB, the S/N drops by 6 dB due to the increase of the quantizing steps; for $V/V_{max} < -18$ dB, the segment slope increases, i.e. the step size decreases by a factor of 2, yielding 6 dB more of S/N, etc. for the adjacent segments. For $V/V_{max} < -42$ dB, the μ-law (μ = 255) provides another 6 dB increase over the A-law due to the factor of 2 higher slope of the segment that crosses zero.

Due to considerable commercial interest in telecommunications, many implementations of coders and decoders, abbreviated as 'codecs', have been published. Some examples are given below to illustrate the state of the art. Most of them can be assigned to one of the three following categories:

(A) Linear converters followed by digital code converters to get 8-bit from originally 12/13-bit linear coding.
(B) Successive approximation, i.e. bit-by-bit conversion based on a non-linear D/A converter, the structure being that of a feedback converter (Fig. 3.25).
(C) Parallel converters for the non-linear spacing of segments combined with interpolation in a feedback loop for the linear subdivision of one segment (Fig. 3.29).

The approaches differ in the technologies (bipolar, NMOS, CMOS) used for the analog and digital circuit sections and in the way the scaling of weights is provided, for instance, as resistor- or capacitor-based divider networks.

(A) Linear coding, digital companding.
The principle used in reference 57 is coarse/fine counting resulting in a linear 13-bit code that is subsequently converted to 8-bit μ-255 or A-87.6 companding law by an external digital code converter. Two chips are used, one bipolar for analog

functions and one NMOS for control logic. Another paper[58] describes a linear delta-modulation circuit with refinements, followed by a delta-to-PCM conversion that feeds into a linear PCM to A-law encoding. It is implemented as a hybrid circuit package. Another two-chip solution (bipolar for analog amplifiers, comparator and reference, MOS for control logic) is proposed in references 59 and 60 using a successive approximation feedback structure with a linear D/A converter and a code converter between the 8-bit successive approximation register and the 12-bit D/A converter.

(B) Successive approximation with non-linear D/A converters.
Most proposals fall into this category because the structure is very efficient from a component-count point of view. Non-linear D/A converters are described in Chapter 4, so these will be excluded here. One chip in NMOS technology is needed for a complete codec.[61] The non-linear D/A conversion is based on a resistor voltage divider with the linear subdivision carried out by unit resistors connected to the non-linear divider taps. A metal gate CMOS one-chip solution is described in reference 62, another one-chip I^2L bipolar approach in reference 63. Another NMOS version includes CCD transversal filters on the chip, and the analog references are obtained by a combination of a linear resistive and a non-linear capacitive voltage divider.[64] A two-chip CMOS codec is described in reference 65 which is based on the capacitive charge redistribution technique. Two capacitive voltage dividers are cascaded to carry out the segment and step decoding. Another two-chip approach[66] uses a D/A converter, where the non-linear segments are produced by tapping an R/2R ladder structure by means of an analog multiplexer. The linear steps are obtained by linear taps of the longitudinal resistor R of the ladder.

(C) Interpolative non-linear converters.
A simple binary weighted network is used in a µ-255 converter to provide the non-linear segment spacing in connection with interpolation for the linear subdivision into sixteen values (an interpolative converter; see section 3.4.4). The encoder[67] is based on bipolar I^2L-technology consisting of 550 gates. In a similar way, the decoder is established on another chip carrying 300 I^2L-gates.[68]

All the versions described above are proposed for use on a per telephone channel basis as opposed to the multiplexed operation of earlier converters (see section 3.6.3). One of the earliest per channel circuits was proposed in 1972.[69] This uses two sample-and-hold circuits interconnected by switches such that, by transferring the signal, it is divided by a factor of 2 each time. The fine linear measurement is done by a linear ramp counting circuit. An analog transfer function based on the folding encoder (section 3.3.4) giving a non-linear piecewise linear characteristic for the A-law is described in reference 30.

3.6.3 Floating Point Analog-to-Digital Converters
Speech coding requires a non-linear encoding characteristic due to its high

dynamic range with a constant signal-to-noise ratio. There are other applications, such as instrumentation and control, with similar requirements, i.e. continuous signals that have limited relative accuracy but vary over several decades. In these cases, a non-linear A/D converter with a logarithmic characteristic provides a satisfactory and economic solution.

There are several ways in which logarithmic conversion can be achieved. These can be grouped into three main classes according to reference 70:

(1) Cascade connection between an analog logarithmic conversion (amplifier or piecewise linear structure) with analog output followed by a linear ADC, as indicated in Fig. 3.37(a).
(2) Linear ADC followed by a digital code converter doing the assignment by a coding table stored in a ROM or by a programmable logic array. This approach is straightforward, but not effective as regards components.
(3) The logarithmic conversion and the A/D conversion are performed by the same structure and cannot be easily distinguished.

The structures are represented in Fig. 3.41. Here we will deal only with the third class of devices, because in this case the savings in word length as compared with linear conversion also directly affect the amount of hardware needed for the converter itself.

True logarithmic conversion is difficult to achieve because the logarithm of

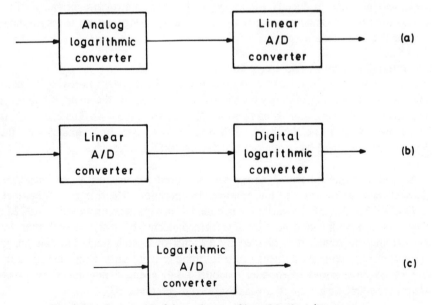

Fig. 3.41 Structures of three classes of logarithmic A/D converters

negative numbers is not defined and $\log 0 = -\infty$. Therefore practical converters are based on a floating point notation, where a number N is defined as

$$N = S_M \times M \times B^E$$

where S_M = Sign of the number N
M = Mantissa or fraction of the number
E = Exponent $(E > 0)$
B = Base of exponent

We will restrict ourselves to $B = 2$. The total word length is then

$$n = b + e + 1$$

where b = number of bits for the mantissa, excluding the sign bit
e = number of bits for the exponent.

There are other floating point representations which allow a larger input dynamic range at a given quantization error when the word length is fixed, but they are usually more complex to implement.

Two important parameters for floating point notation are dynamic range and relative error. If the mantissa is normalized to values below 1 with a maximum of $(1 - 2^{-b})$, the smallest and largest number $|N|_{\min}$ and $|N|_{\max}$, respectively, are

$$|N|_{\min} = \frac{1}{2^b} \cdot 2^0, \ |N|_{\max} = \left(1 - \frac{1}{2^b}\right) \cdot 2^{(2^e - 1)}$$

So the dynamic range D is

$$D = \frac{|N|_{\max}}{|N|_{\min}} = (2^b - 1) \cdot 2^{(2^e - 1)} \approx 2^{(b + 2^e - 1)}$$

D is mainly determined by the length of the exponent. The maximum relative error $\varepsilon(N)_{\max}$ is

$$\varepsilon(N)_{\max} = (2^b + 1)^{-1}$$

the relative error being defined as

$$\varepsilon(F) = \frac{|F - N(F)|}{|F|}$$

where F as well as $\varepsilon(F)$ represent continuously varying figures and $N(F)$ the discrete representation within our floating point number system. Excluding values below the smallest $N(F)$, the maximum is in between two possible values for

$$F = \frac{N_i + N_{i+1}}{2}$$

yielding $\varepsilon(N)_{max}$ as maximum between two adjacent zeros at the values N_i and N_{i+1}, respectively.

Let us illustrate the situation by some comparisons between floating and fixed point notation representations. Assuming a floating point number of $n = 8$ bit length, where $e = 3$ and $b = 4$ and one bit for the sign, then the dynamic range is

$$D = (2^4 - 1) \times 2^7 = 1920$$

For the same dynamic range, 11 bits would be needed in a fixed point representation. For a linear converter (one bit also for the sign)

$$D = (2^7 - 1) = 127$$

The maximum relative error for floating point is

$$\varepsilon(N)_{max} = (2^4 + 1)^{-1} \approx 0.06\%$$

whereas for fixed point or linear representation at full-scale value N_{max}
$\varepsilon(N)_{max} = (2^8 + 1)^{-1} \approx 0.004\%$.

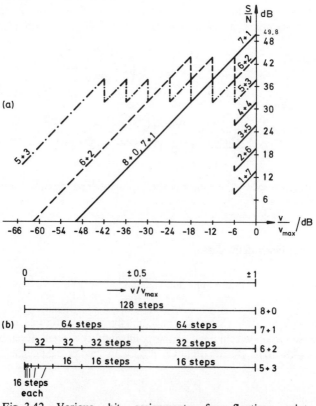

Fig. 3.42 Various bit assignments for floating point representation for $n = 8$. (a) Signal-to-noise ratio versus signal level, log scale; (b) subdivision of input signal range, linear scale

A measure for the relative error is also the signal-to-noise ratio. Let us again take our example for $n = 8$, i.e. 1 sign bit and 7 bits for the magnitude. These 7 bits can be assigned to mantissa and exponent in various ways. The signal-to-noise ratio S/N versus signal amplitude can be plotted each time; for instance, for a sine wave signal. This is done in Fig. 3.42 for some examples.

The maximum value for S/N at $V/V_{max} = 1$ is obtained for $(8 + 0)$, i.e. no bit for the exponent. In this case of uniform quantizing, all steps are equal in size. So the full-scale relative error is lowest and S/N exhibits its maximum at full scale. Assigning 1 bit to the exponent does neither good nor harm, because each half of the range is coded with 64 steps as in the case of $(8 + 0)$. The assignment $(6 + 2)$ gives 32 steps for the upper half of the range, and the maximum S/N value correspondingly drops to 37.8 dB. Thirty-two steps fall within $1/4 < V/V_{max} < 1/2$, as in the $(7 + 1)$ case, with 32 steps also between $1/8 < V/V_{max} < 1/4$ and 32 steps between $0 < V/V_{max} < 1/8$, yielding an improvement of 6 dB and 12 dB over the $(7 + 1)$ case, respectively. For $(5 + 3)$ the maximum drops again by 6 dB, because only 16 steps are within the upper half; the advantage is seen at lower levels due to the 8 segments of the idealized picture. The $(5 + 3)$ corresponds to the $\mu = 255$ law shown in Fig. 3.38. Figure 3.42 shows the trade-off between maximum signal-to-noise ratio and dynamic range as a specific characteristic of floating point conversion.

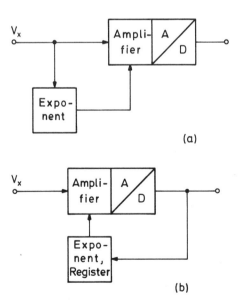

Fig. 3.43 Determination of exponent and mantissa for floating point A/D conversion. (a) Exponent derived by additional hardware; (b) A/D converter determines both exponent and mantissa

Table 3.5 Analog input and corresponding digital output of a 3-bit ADC

Analog value V	MSB b_2	NMSB b_1	LSB b_0
7	1	1	1
6	1	1	0
5	1	0	1
4	1	0	0
3	0	1	1
2	0	1	0
1	0	0	1
0	0	0	0

As previously stated, we will cover here only the third class of converters of Fig. 3.41 where the converter itself is logarithmic. This can again be done in two ways using a linear converter and a variable gain amplifier. Consider Fig. 3.43(a), where one converter block determines the exponent or range in which the signal is to be measured. This then modifies the gain of the amplifier such that the linear ADC is used to full scale.

More efficient is the structure of Fig. 3.43(b), where the linear ADC is used twice. First, it obtained the measurement for the exponent with the amplifier gain at its lowest value. Second, the gain of the amplifier is set such that the ADC is used to full scale. In the third and final step, the mantissa is obtained. Instead of

Table 3.6 Amplifier gain depending on the digital output of the first measurement (Fig. 3.43(b))

Digital output	Gain A	Exponent
$b_{n-1} = 1$	1	0
$b_{n-1} = 0$ $b_{n-2} = 1$	2	1
$b_{n-1} = 0$ $b_{n-1} = 0$ $b_{n-3} = 1$	4	2
$b_{n-1} = 0$ $b_{n-2} = 0$ $\vdots$ $b_{n-(k-1)} = 0$ $b_{n-k} = 1$	2^{k-1}	$k-1$

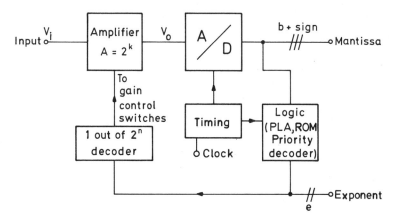

Fig. 3.44 Floating point A/D converter according to Fig. 3.45(b)

using an amplifier with variable gain, a D/A converter can be used to change the reference voltages of the ADC according to the first measurement.

Being used twice, the resolution of the linear ADC restricts the number of bits for the exponent. If it has n unipolar bits, then $2^n \geqslant N_{max} \approx 2(2^e - 1)$ for determining the gain within one step, that is, $n \geqslant 2^e - 1$. Otherwise, using i steps, $i \cdot n \geqslant 2^e - 1$.

Let us consider an example of an unipolar linear ADC with $n = 3$ bit which in one step can be used for an exponent of $e = 2$ bit length. In Table 3.5, the analog values as well as the corresponding eight digital words are listed for the straight binary code. For full-scale resolution of the ADC for the mantissa, the MSB is $b_{n-1} = 1$; the amplifier gain is then $A = 1$ and the corresponding exponent is 0. The analog values $3 > V > a$ have to be amplified for full-scale resolution by a factor of 2, the exponent gets 1, etc. Following this procedure, we obtain Table 3.6.

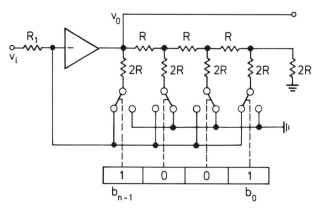

Fig. 3.45 Variable gain amplifier by D/A converter in a feedback configuration

From Table 3.6 it follows that the leading '1' has to be determined in the digital output word of the first measurement. This can be done by a table stored in a read-only memory (ROM), a programmable logic array (PLA), or by a priority decoder. The basic structure of the floating point converter of this type is shown in Fig. 3.44.[71] The value of the exponent is stored at the same time in a register that controls the gain of the exponential gain amplifier. Such an amplifier can be implemented as shown in Fig. 3.45. It consists of an operational amplifier that has a R–2R ladder network in its feedback path. As described in the section on D/A converters, the gain is therefore

$$A = \frac{V_0}{V_i} = \frac{2^n R}{R_1} \cdot \frac{1}{\sum_{k=0}^{e-1} b_k 2^k}$$

Instead of an implementation based on medium-scale integrated circuit logic building blocks, a microcomputer can be used to take over the functions of timing, logic, and gain control.[72] The hardware, in addition to the microcomputer, can be considerably reduced to a comparator C, two multiplying D/A converters, and two operational amplifiers when using the successive approximation algorithm for A/D conversion. The structure is shown in Fig. 3.46. For a 6502 microcomputer with a clock frequency of 1 MHz, the program length is 100 bytes and the time needed for one floating point A/D conversion is 1.5 msec.

A diagram of a characteristic of a floating point converter with an exponent length $e = 3$ bit, $b = 7$ bit for the mantissa and 1 bit for the sign in bipolar operation can be displayed on an oscilloscope by connecting the analog input voltage to the horizontal deflection and converting the mantissa back to analog by a suitable D/A converter. This is shown in Fig. 3.47.

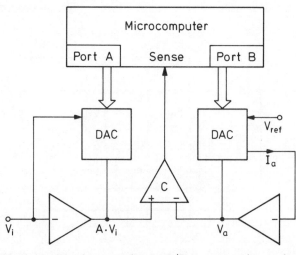

Fig. 3.46 Floating point A/D conversion by microcomputer control

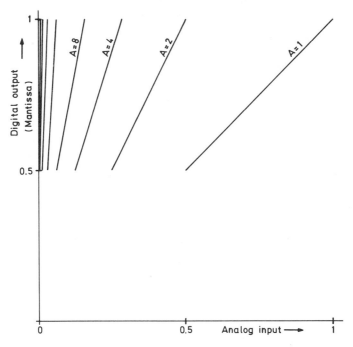

Fig. 3.47 Transfer characteristic of a floating point A/D converter for $e = 3$

3.6.4 Multiplexed Analog-to-Digital Converters

Introducing digital methods and/or circuitry in practical systems is primarily a question of the associated costs. Special consideration must be given to the additional cost of the A/D converters. There are many applications in which a single converter can be time-shared for more than one signal source and its cost be shared accordingly. Examples of these applications are process control, telemetry, and PCM-telephony. Due to the fact that the cost of A/D conversion increases sharply only beyond a certain speed limit, conversion cost per signal source (channel) reduces appreciably by increasing the number of sources shared by the converter when staying below the speed limit.

A block diagram of a system that shares a single A/D converter in time is illustrated in Fig. 3.48. As shown, each signal is first low-pass filtered before it is applied to an S/H circuit. The timing of the S/H process is cyclically controlled by a synchronizing circuit. The resulting amplitude-modulated train of pulses is then applied to the A/D converter. In order to avoid crosstalk between the individual channels the S/H circuit must meet stringent requirements. For this purpose the time interval allocated for a single conversion in the digitizer should be smaller than the sampling period divided by the number of inputs.

After processing (in a data acquisition system) or at the receiving end (in a communications system), D/A conversion is first carried out and then followed by

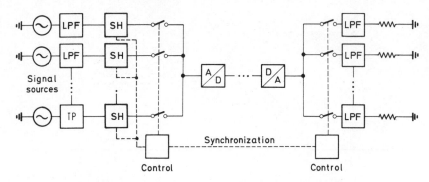

Fig. 3.48 Block diagram of a time shared data converter

a synchronized redistribution to the individual channels (subscribers). Through appropriate synchronization between sending and receiving ends, it is guaranteed that each source is connected to its proper sink. If connections between source and load (subscriber) can be selected through any form of control and addressing, then such a system can provide the switching function of a telephone exchange. Such 'electronic' exchanges are increasingly used. In this case, the operation could be controlled and monitored by a centralized computer, the stored program of which can easily be changed.

The PCM-system introduced by the ATT-company (T1-system) is designed for twenty-four telephone channels and uses a single A/D converter for twelve subscribers. In Europe, the recommended number of channels by the CEPT ranges between twenty-four and thirty-two subscribers. A single 8-bit digitizer is used at a sampling rate of 8 kHz (i.e. a sampling period of 125 µs). The resulting bit rate (32 channels) is 2.048 Mbit/s.

3.6.5 Delta Modulators

Delta modulation represents another alternative for an economical A/D conversion technique.[73,74] Savings on cost are achieved by extreme simplicity of the hardware required. This circuit simplification is due to the fact that only the difference in amplitude between every two succeeding samples, rather than their absolute values, is coded with a single bit. This is an extreme case of DPCM encoding, as described in section 2.2.

The converter is accordingly very simple and consists, as shown in Fig. 3.49(a), of only a comparator, an AND-gate, and an approximation network (predictor). The predictor is required to deliver an estimation (prediction) of the signal that depends on a pulse train already transmitted. At the receiving end, the circuitry is even simpler, and consists of a predictor as in the transmitter and a low-pass filter. As shown in Fig. 3.49(a), the comparator decides whether a clock pulse is to be transmitted via the AND-gate (signal > predicted value) or not (signal < predicted value).

The predicted values are represented in Fig. 3.49(b), for simplicity, through a

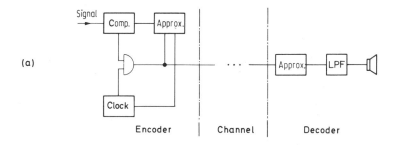

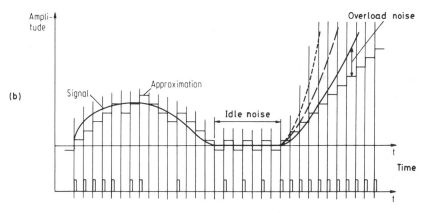

Fig. 3.49 Delta modulation. (a) Block diagram of a simple delta modulator; (b) approximating the signal through a step function

step function. Actually, the approximating function is composed of a combination of the first and second integration of the step function. As a result of this approximate reconstruction of the signal, the quantization leads to audible errors at those periods where the signal is zero or undergoes a sudden change. In the first case, the resulting noise is called the 'idle or granular noise', which depends on the height of the step. Trials to reduce the step size, in order to minimize this noise, lead to a situation where fast variations in the signal cannot be followed by the approximating function. Such noise is called 'overload noise'. Therefore one has to make a compromise, concerning the step size, between idle and overload noise. An attractive approach is to use some sort of 'adaptive' step height according to the signal conditions, as is the case in adaptive or companded delta modulation. The converter accordingly increases in complexity.

The clock frequency is another important parameter in such a converter. In the simple delta modulation for speech, the clock frequency and, correspondingly, the required channel capacity is about 50% larger than that with PCM (100 kbit/s compared with 64 kbit/s for the latter). However, by using companded delta modulators the same quality as in PCM can be obtained at a rate of only 32 kbit/s.[74]

Delta modulation is especially suitable for speech communication, since the signal power spectrum drops sharply at high frequencies. Although the spectrum of video signals also reduces at high frequencies, the technique was found to be less suitable for video encoding, since the eye is very sensitive to fast brightness transitions that are affected by overload errors. Delta modulators are said to be sensitive to transmission errors. Several improvements such as 'leaky integration' are described in the references mentioned above. It is to be noted here that delta modulators provide no binary coding.

3.6.6 Stochastic Analog-to-Digital Converters

Due to their very simple architecture, stochastic A/D converters provide economical realizations and high noise immunity at the expense of an increased conversion time, so their use is limited to low-frequency applications. The code representing the signal is not the straightforward binary one, but rather has the form of a seemingly random pulse train. Measuring equipment that can deal with stochastic coding is available for some applications.[75]

Let us assume that the input voltage $V(t)$ to be digitized is a stationary function, e.g. in the simplest case, a d.c.-voltage with a value $V(t) = V_0$, and that the probability p of a certain event can be represented by counting its frequency of occurrence.

As shown in Fig. 3.50, a stochastic A/D converter consists of a comparator and a threshold-voltage generator whose output $r(t)$ together with $V(t)$ are the comparator's inputs. The output is

$$z(t) = v_1 \doteq \text{'1' if } V(t) > r(t)$$

If $r(t)$ is a signal whose amplitude is equally distributed over the full-scale range of $V(t)$, e.g. to simplify, a sawtooth voltage that swings the FSR of $V(t)$ in the period t_p, then from Fig. 3.51, due to geometric considerations we get

$$t_0/t_p = V_0/V_{FS}$$

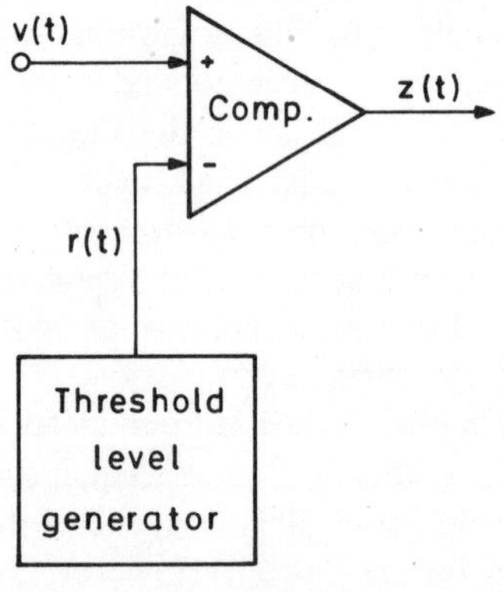

Fig. 3.50 Configuration of a stochastic converter

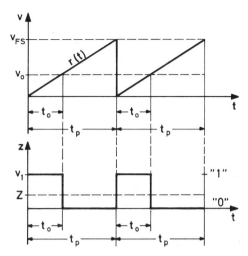

Fig. 3.51 Waveforms corresponding to Fig. 3.52

In other words, the ratio t_0/t_p is proportional to the input voltage V_0. The mean value of $z(t)$ is given accordingly by

$$Z = \frac{1}{t_p} \int_0^{t_p} z(t) dt$$

leading to $Z = V_0$ for $V_1 = V_{FS}$. If the comparison of $V(t)$ with $r(t)$ is frequently repeated and the resulting signal $z(t)$ is statistically sampled, then the probability

$$p\{z(t) = V_1 \triangleq \text{'1'}\} = t_0/t_p$$

i.e. it is also given by the ratio t_0/t_p. This conclusion remains valid if the sawtooth voltage is replaced by any random waveform generator that has a constant probability density distribution for all amplitudes within the FSR.

The probability $p\{z(t) \triangleq \text{'1'}\}$ therefore represents the coded value of the analog quantity V_0. The output $z(t)$ is a sequence of '1' and '0', i.e. it is a digital representation of V_0 which can be transmitted and/or processed in this form. Obviously, it is very insensitive to interfering pulses as long as their number remains sufficiently small.

The determination of V_0 by means of the output $z(t)$ has to be carried out over a finite time interval. The resulting values will accordingly have a certain standard deviation

$$\sigma\left\{\frac{Z_r}{Z}\right\} = \sqrt{\left(\frac{1-p}{p} \cdot \frac{t_c}{2T}\right)}$$

where Z_T is the actual mean value measured within the time interval $-T < t < T$ whereas Z is the true mean value.[75] The time t_c is the clock period, assuming that $z(t)$ is a sampled pulse train, so that $2T \cdot p/t_c$ is the average number of pulses during observation time $2T$.

It is evident that in order to obtain smaller variances, i.e. higher precision, the measuring interval $2T$ must be large and must include a correspondingly large number of periods t_p, t_p being the time needed by the random function generator to scan the FSR once.

Another attractive property of this conversion technique is the fact that the corresponding decoders (D/A converters) are even simpler than the encoder. A stochastic D/A converter is obtained by a circuit forming the mean value of $z(t)$. This function can be provided by a low-pass filter, e.g. by a simple RC network (Fig. 3.52). In this case, the time constant $\tau = RC$ must be chosen large enough to provide the required precision (low ripple). For a permissible error of $\pm 1/2Q$ in an equivalent n-bit binary converter, the time constant is given by

$$\tau = 0.35 \cdot t_i \cdot n \cdot 2^n$$

where t_i is the duration of one pulse of the statistical pulse train.[76] Moreover, the time t_m to reach steady state must include several time constants, i.e.

$$e^{-t_m/\tau} \leqslant \frac{1}{2^{n+1}}$$

For a clock frequency of 1 MHz, i.e. $t_p = 1\ \mu s$, the time t_m for $n = 10$ bit is 27 ms while for $n = 8$ it is only 4.5 ms. Applying the sampling theorem, we can conclude, using the above values, that for a resolution of $n = 10$ bit only frequencies of up to 18 Hz and with $n = 8$ bit frequencies of up to 110 Hz can be converted. These values can be improved by about one order of magnitude if a second-order low-pass filter is used.

A further interesting method is to reconstruct analog signals from their binary-coded representation using stochastic coding as an intermediate step by using standard digital integrated circuits. As shown in Fig. 3.53, the binary coded input is temporarily stored in a register and then compared in a digital comparator with the output of a random-number generator.[77] Since it is required only that all the produced values should be equally probable, then an up/down counter, or even a

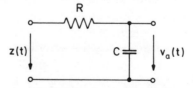

Fig. 3.52 RC low pass as a simple stochastic D/A converter

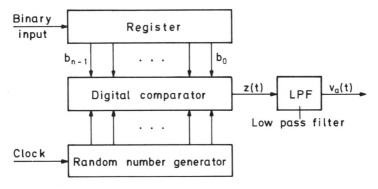

Fig. 3.53 D/A converter for *n*-bit binary coded inputs using the stochastic principle as an intermediate step

ripple counter, for a corresponding range of numbers can be used. The comparator output $z(t)$, the stochastic coded value, is then low-pass filtered to yield the reconstructed analog signal $V_a(t)$. In this method, no precise current sources or resistors are needed as is usually the case in conventional D/A converters (Chapter 4). An A/D converter can be obtained if a D/A converter of this type is inserted in the feedback loop of a cyclic A/D converter (Fig. 3.25). However, the required conversion time is relatively large.

As pointed out above, the use of the stochastic technique is limited to low-frequency signals. However, a similar technique was successfully used to encode video signals in the so-called 'dithering' technique. From a two-level pulse train, the grey levels of a picture are generated via the monitor and the human eye to produce 'subjectively' the same effect as an analog signal.[78] If the threshold levels of a low-resolution A/D converter, 4 bits for example, were varied through a superimposed 'dithering' voltage, then contour effects which are due to low-amplitude resolution can be avoided or softened, respectively.

3.6.7 Ultra-High-Speed Analog-to-Digital Conversion

Ultra-high-speed converters in this context are considered to be those working beyond a 100-MHz word rate. Most of the converter approaches in this speed range are at a more or less exploratory stage of development, i.e. their parameters are preliminary or are extrapolated from one-stage experiments or are even paper estimations only. This should be kept in mind when quantitative parameters are quoted. Nevertheless, this stage of exploration is necessary to bring the state-of-the-art towards Gigahertz sampling rates.

As reported in section 3.3, the maximum speed was a 100 MHz sampling rate at 8-bits resolution.[20] It is a stretched version of a 20 MHz, 10-bit experimental converter, based on a sequential–parallel structure with pipeline operation. (Pipelining is explained in section 3.3.5.) This is useful in high-speed operation to trade delay for resolution. Without pipelining, the parallel structure based on

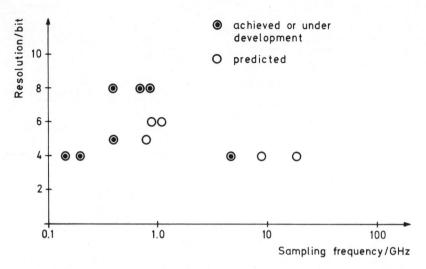

Fig. 3.54 State of the art in A/D converter development as of mid-1976[81]

silicon ECL circuit technology is limited to a 150 MHz word rate for 6 bits[9] or 350 MHz at 4 bits.[79]

In 1968 a 240 MHz word rate, 5 bits per sample converter was described,[80] based on a microwave carrier phase-shifting technique. More recent approaches concentrate on five areas:

(1) Extension of the electron beam semiconductor target (EBS) encoding tube (see section 3.2.1).
(2) Transferred electron devices (TELD) and GaAs field effect transistors (see section 5.2.5).
(3) Josephson junction (JJ) as high-speed comparators (see section 5.3.5).
(4) Electro-optic or acousto-optic deflection of a light beam.[81,93]
(5) Transient waveform storage with subsequent conversion (see section 3.6.8).

As it would not be appropriate in this book to describe the details of the devices being used an overview article is recommended.[81] Fig. 3.54, summarizing the results obtained so far, is taken from that reference.

3.6.8 Transient Analog-to-Digital Conversion

Transient converters are used in experiments or supervision equipment where a burst of samples occurs in an event producing fast transient waveforms that need high resolution in time during a relatively short interval. The burst of samples is taken in real time into a shift register and stored in intermediate form for lower-rate A/D conversion afterwards.

Charge-coupled devices (CCDs)[82] are widely used as a buffer memory for word

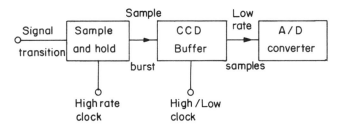

Fig. 3.55 Charge coupled device as a word rate translator

rate translation according to the structure of Fig. 3.55. Sample bursts of up to 250 MHz rate have been used. If this is not sufficient, interleaving is introduced, often up to four times in order to achieve a 1 GHz rate.[83] According to reference 84, 1 GHz CCD rates should be achieved in silicon, whereas GaAs should allow even a 5 GHz rate. If the length of one CCD is not sufficient, a series/parallel/series (SPS) structure can be applied,[85] as indicated in Fig. 3.56, to overcome the high-frequency attenuation within a CCD.

Sampling rates up to 1 GHz can also be obtained by using a stripline sampling technique.[86] Here the signal travels on one stripline and the sampling pulse on another. At regular intervals a GaAs field effect transistor operates as a sampling

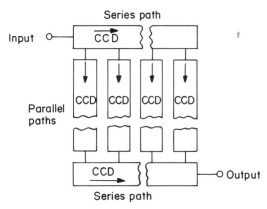

Fig. 3.56 Series–parallel–series operation of multiple CCDs

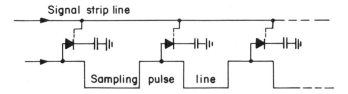

Fig. 3.57 Principle of stripline sampler

gate by connecting its gate electrode to the sampling line, its source electrode to the signal line, and its drain to a storage capacitor. Due to unequal length of the line segments (Fig. 3.57), the sampling pulse shifts with respect to the signal. An experimental unit of sixteen sections has been tested at a 1 GHz rate, and 512 sample values seem feasible. The rate after time translation can be as moderate as about 100 kHz in order to allow high-resolution, low-cost A/D converters to be used.

3.7 Analog-to-Digital Conversion and Microcomputers

3.7.1 Introduction

As pointed out in Chapter 1, A/D converters are interfaces between the real analog world and digital processing, storage, or transmission. Therefore in areas such as instrumentation and process control, A/D conversion has to be carried out on computers. This is now considered to be a conventional or standard type of application. It is characterized by a loose coupling between the A/D converter and the computer, i.e. there is little or no interaction which is related to the A/D conversion itself.

However, since the advent of microprocessors and such powerful technologies as LSI or VLSI much more interaction between the microcomputer and the A/D converter can be observed due to their proximity as semiconductor circuit chips. In fact, a number of functions to be carried out for A/D conversion is taken over by the microcomputer itself, giving a new dimension to the A/D conversion process, i.e. that of controlling A/D conversion by software.

At present, high-performance converters require precise linear circuits working at small voltage levels and up to high frequencies and precisely controlled supply voltages that are stable and free of noise. Therefore separate chips for analog and digital circuitry, i.e. hybrid solutions, are widely used. The need for low cost and high volume tends to bring both types of circuits on to one chip. Data acquisition building blocks based on either CMOS or bipolar technologies already provide many functions such as analog multiplexer, sample-and-hold, comparator, address decode, control logic, and intermediate storage.[87] Special purpose analog signal processing such as digital filtering is also available as a monolithic circuit.[88]

The future will see much more sophistication on one chip, where the A/D converter is only one function out of many, yielding an increased proliferation of digital signal handling.

3.7.2 Analog-to-Digital Conversion by Software

In this section we do not imply A/D conversion by software only but stress the point, related to microcomputer-based structures, that the same hardware can be used to implement various algorithms.

Consider Fig. 3.58, where a standard microcomputer containing a central processing unit (CPU), a random access memory (RAM), a read-only memory

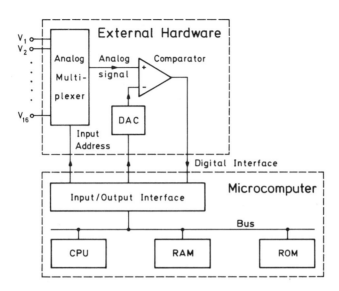

Fig. 3.58 Microcomputer structure extended by external hardware for A/D conversion via standard input/output interface

(ROM), and an input/output building block is shown together with some additional circuitry for multiple input A/D conversion. The ROM stores the software, i.e. the program according to which the conversion is carried out. The read/write memory (RAM) is used to store the results of the A/D conversion. The central processing unit organizes the co-ordination of all units in the system. It addresses the multiplexer, offers an initial or improved estimation value for comparison through the digital-to-analog converter, and receives the result of the comparison for making decisions on how to proceed with the conversion process.

The type of conversion algorithm depends on the program stored in the read-only memory. Figure 3.59 shows a flow diagram for the counting method, Fig. 3.60 a straightforward method for the successive approximation algorithm. Excluding the additional commands for initialization needed for setting the input/output building block according to the intended operation mode, for saving the status and information of the CPU registers the counting algorithm takes about 25 bytes of program, and for the successive approximation about 35 bytes are needed, based on the instruction set of the 8080/8085 microcomputer.

The time needed to carry out one single conversion for 8-bit resolution varies between about 100 machine cycles and about 18 000, depending on the magnitude of the analog sample, for the counting method. Taking again the 8085 machine cycles of 325 ns as an example, the conversion time varies between about 33 µs and about 6 ms, i.e. the bandwidth of an analog signal that can be encoded is below 100 Hz.

The successive approximation technique requires the number of cycles of the inner program loop to equal the number of bits n. For $n = 8$, about 800 machine

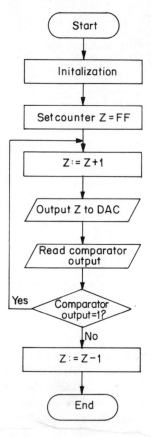

Fig. 3.59 Flow diagram for counting algorithm (8-bit)

cycles of 325 ns are needed, leading to a conversion time around 250 µs and a maximum analog signal bandwidth of about 2 kHz, respectively. Similar results are obtained for the 6809 microprocessor.

The maximum signal bandwidth quoted for the examples are valid only if one input of the multiplexer is used. They become correspondingly smaller when the number of inputs is increased. In addition, time is needed to address the multiplexer, and this further decreases the highest conversion rate.

For many applications such as heat control, supervision of environmental conditions, or slow-motion mechanics, this level of speed is sufficient. It is even fast enough to leave the processor a considerable amount of time idle if not used by other functions. However, for fast-moving motors and/or multiple signal conversion in connection with preprocessing for data compression, it is desirable to improve the speed of conversion.

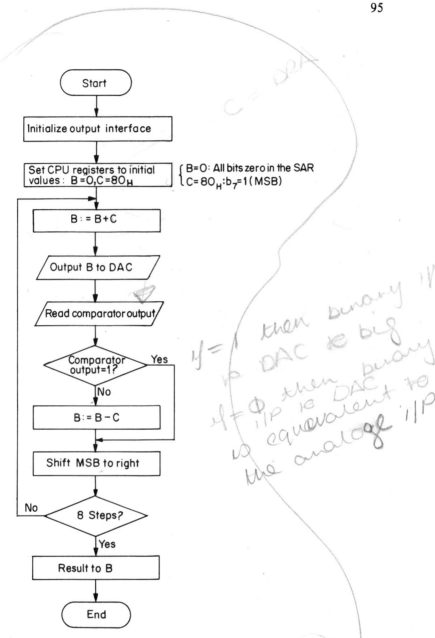

Fig. 3.60 Flow diagram for successive approximation algorithm (8-bit)

3.7.3 Data Acquisition Building Blocks

By adding some hardware to the peripheral building block the microcomputer can be considerably less loaded by the conversion itself. One common method of doing this is shown in Fig. 3.61, where a successive approximation register (SAR) and a hardwired control logic (CL) have been added. Upon a request by the microcomputer, the CL starts and completes a conversion cycle for one input, the address of which is stored in the address register. During the conversion time, the microcomputer is free for other uses.

After the conversion, the digital sample value is stored in the SAR. This can be indicated by the CL, causing the microcomputer to interrupt its operation in order to receive the sample value for storing it into its RAM. Due to the fact that the microcomputer knows in advance how long the conversion takes, the microcomputer can itself stop its other activity by means of an instruction in its program (polling or software interrupt) and pick the sample value out of the SAR.

Figure 3.61 is intended to show, in addition, another aspect. By comparing this figure with Fig. 3.58 it can be seen that the input/output building block has been removed and that the microcomputer bus is shown in more detail, consisting of address, data, and control lines. In other words, the data acquisition building block is directly connected to the MC bus. By this means it can be addressed in the same way as the RAM and ROM blocks, i.e. the external data acquisition unit also shares the address space of the MC. This mode of connection and organization is called 'memory mapped input/output'.

By this means, no input/output instructions, which usually take time to be carried out, are used. Instead, the faster instructions for memory access operations are employed. The only drawback here is that some addresses are then

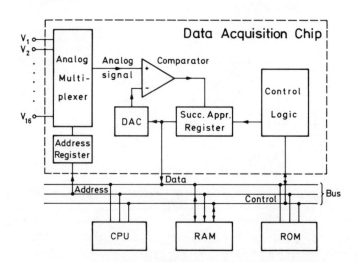

Fig. 3.61 Microcomputer structure extended by data acquisition chip for A/D conversion via memory mapped input/output

occupied by input/output devices, leaving less address space for RAM/ROM extensions.

The increase in speed can be considerable. Depending on the technology of the data acquisition chip, the time for one conversion can be reduced to several microseconds, grading up the bandwidth limitation to the hundreds of kilohertz range.

Of course, the data acquisition building block can also be connected to the MC via an input/output building block, as indicated in Fig. 3.58. Even more flexibility is offered to the designer by using bit slice processor building blocks. For further details the reader is referred to the literature on microcomputers.[89-91]

3.7.4 Vertical Analog-to-Digital Conversion

In the case of several inputs another method of increasing the speed per input is to interleave the A/D conversion in time in a way different from that assumed for Fig. 3.58, where all the inputs are treated sequentially. Thus the conversion of input k is completed before the conversion of input $(k + 1)$ is achieved (Fig. 3.62(a)). This is very inefficient from the point of view of the microcomputer, because only one bit of its n-bit word length is used at a time (Fig. 3.62(b)). A more efficient way of interleaving that uses the whole length of the microcomputer word is shown in Fig. 3.62(c), where the MSBs of k inputs are handled at a time in parallel, filling the whole word of $n = k$ positions. The next less significant bits are obtained correspondingly. After n conversions, all $k = n$ inputs are served. The bits belonging to one input are located vertically above each other in the microcomputer memory, thus explaining the name of this technique.[72]

Obviously, the external hardware has to be modified to carry out this operation. Also, in order to avoid losing some of the time saved by rearranging the words within the microcomputer before further processing it is advisable to use a bank of SAR registers that are used as serial/parallel converters when transferring the words into the microcomputer memory. A block diagram of the external circuitry is shown in Fig. 3.63 for four inputs, where the input/output building blocks part A (PA) and part B (PB) are used as unidirectional outputs and part C (PC) as a unidirectional input.

A more efficient solution can be obtained by using the registers within the input/output interface building block for both input/output and conversion of rows and columns.[71] In a similar way, D/A conversion for $k = n$ outputs can be obtained.

3.8 Summary and Conclusions

Precision, speed, and hardware complexity represent the main co-ordinates of an A/D converter. Usually, precision and speed are predetermined by the specific application. This means that the admissible quantization noise, on the one hand, and the maximum input-signal frequency, on the other, are first-priority design

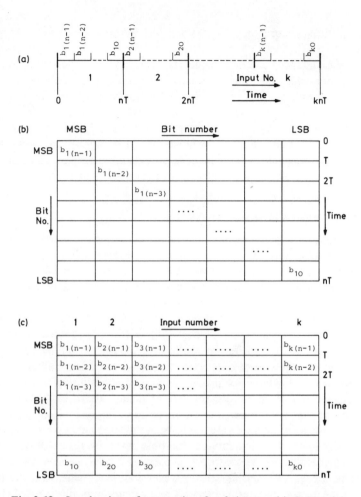

Fig. 3.62 Interleaving of conversion for k inputs. (a) Sequential conversion for k inputs of n bits each; (b) subdivision of conversion for input 1 into n intervals; (c) vertical interleaving of k conversions within $n = k$ intervals

factors. Complexity of conversion hardware is a qualitative rather than quantitative aspect and, therefore, it is difficult to evaluate. However, in this respect, two current tendencies can be observed:

(1) Within the converter itself the trend is to shift the interface between the analog and digital circuitry in favour of the latter.
(2) Conversion techniques and circuit implementations, which can be produced in a monolithic integrated-circuit form, are strongly favoured. They will provide robust, economical, and reliable converters in the near future.

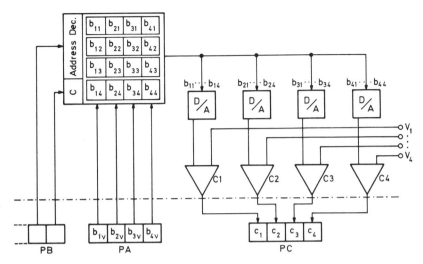

Fig. 3.63 Circuit block diagram for vertical A/D conversion using external registers for row/column interchange of four inputs[41]

Examples of such techniques are those which do not imply high-precision components, indirect converters which employ a convenient intermediate quantity (e.g. charge or time) in performing the conversion, and, finally, converters which deliver codes other than the binary code.

The number of conversion steps represent a relative measure for conversion speed. Circuit complexity and/or cost are measured, on the other hand, by the number of comparators required. In Table 3.7 the three different conversion techniques are compared for the case of 10-bit resolution.

From the discussion in this chapter it is clear that the put-and-take technique ranks midway between the extended-parallel and the counting method. It represents a reasonable trade-off between cost and speed requirements. However, it is not optimal whenever the product $N \cdot k$ should be a minimum. The minimum of $N \cdot k$ lies within the extended-parallel method. For the example $n = 10$ of Table 3.7, the minimum is $N \cdot k = 75$, i.e. $N = 15$, $k = 5$. This is always achieved for a

Table 3.7 Comparing conversion techniques (10-bit resolution)

	Parallel method	Put and take technique	Counting method
Number N of references (comparators)	1023	10	1
Number k of conversion steps	1	10	1023
Product $N \cdot k$	1023	100	1023

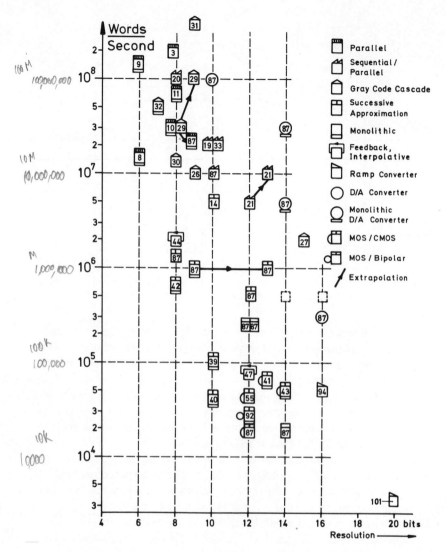

Fig. 3.64 State of the art survey of A/D converters by the end of 1981 (the numbers inside the boxes are those of the references given at the end of this section)

solution with 2 bit per step, which can be implemented in a successive approximation or a cascade structure.

Although the state of the art is moving rapidly, it might be interesting to give a survey representing a selection of examples quoted in this book and from selected sources.[87] Figure 3.64 shows speed in terms of word rate and resolution in terms of the number of bits as the two main co-ordinates. Independent of the techniques and technologies used, the state of the art by the end of 1981 can be

Table 3.8 Classification of the different types of A/D converters[1]

Direct A/D converters

- **No feedback**
  - Pure electronic: **Parallel ADC**
  - Not pure electronic: **Encoder**
- **Feedback**
  - **Not bit-by-bit**
    - **Servo ADC (1 counter)**
    - **Servo ADC (more than 1 counter)**
  - **Bit-by-bit**
    - **1 bit per step sequential**
      - Pre-subtractive: **Successive approximation ADC**
      - Post-subtractive
        - Synchronous
          - Static temporary storage only
          - Dynamic temporary storage: **Cyclic ADC**
        - Asynchronous
          - Static temporary storage only: **Cascaded ADC**
    - **n_k bit per step (sequential–parallel) Post-subtractive only**
      - Synchronous
        - Static temporary storage: **Successive approximation ADC**
        - Dynamic temporary storage: **Cyclic ADC**
      - Asynchronous
        - Static temporary storage only: **Cascaded ADC**

Indirect A/D converters

- **No feedback**
  - **Intermediate**
    - **Time interval**
      - Single ramp method: **Single-slope ADC**
      - Multiple ramp method: **Dual-slope ADC (Triple-slope ADC, etc.)**
    - **Frequency**: **Voltage to frequency ADC**
- **Feedback**: **Combined systems using voltage/time and voltage/frequency conversion**

approximately characterized by a speed × resolution = 10^7 words/s × 10 bit, with a slope of 2 bit/decade increase/decrease.

Other parameters to be taken into account are the methods and technologies used as well as an indication of monolithic implementation. The leading edge performance is still the domain of hybrid converters. Monolithic solutions are limited to 9 bit in the high-speed range and to 14 bits at below 10^5 words/s, i.e. they are 2 bits or a factor of 10 in speed behind hybrids. It would be challenging to establish cost as a third co-ordinate.

At high speed, the sequential–parallel approaches are competing with Gray code solutions. In the medium-speed range, bit by bit successive approximation is more effective and still sufficiently fast. At lower speeds, especially those not shown in the picture ($<10^3$ words/s), ramp converters become predominant, either multiple slope or charge balancing.

Pure MOS converters are exceptions rather than the rule. In the high-speed/high-resolution field bipolars are again leading by 2 bit or one order of magnitude in speed. The medium-speed field contains successive approximation bipolar (I^2L) monolithic converters. Only below 10^5 words/s pure MOS solutions are clustering around 12 bit to 14 bit. As an example for a classification scheme that is different from the one used throughout Chapter 3, consider Table 3.8, which is based on criteria of morphology that are mutually exclusive.[1]

The first criterion is 'direct' versus 'indirect'. The second criterion separates converters with and without feedback. Efficient conversion cannot be achieved without feedback. Two types are essential: 'feedforward' is equivalent to 'post-subtractive', i.e. all pipelining structures fall into that category, whereas 'feedback' in the original sense is equivalent to 'pre-subtractive'. Further criteria are 'bit by bit' or 'not bit by bit' and 'synchronous' and 'asynchronous'.

These criteria have been established as a systematic method of finding undiscovered 'boxes'. At the same time, it limits the scope to existing parameters, whereas new solutions that become attractive through the progress of technology (for instance, charge redistribution or new ways of coding that are different from straight binary or Gray code) are not included in this classification. Therefore the classification followed in Chapter 3 is based on more immediate and illustrative parameters, such as converter architecture, with the purpose of admitting as much freedom as possible to design ingenuity, even at the expense of needing a category for 'miscellaneous' converters.

References

1. Best, R. E., Eine Systemtheorie der DA- und AD-Converter und ihre Anwendung auf die Konstruktion schneller AD-Converter. PhD Thesis, No. 4785, Eidgen. Techn. Hochschule, Zurich (1971).
2. Cooper, H. G., Crowell, M. H., and Maggs, C., 'A High-speed PCM Coding Tube', *Bell Laboratories Record* **48**, 267–72, September (1964).
3. Hayes, R., 'Electron Beam Solves Problems of High-speed Digitizing', *Industrial Research & Development* 124–30, May (1980).

4. Ochs, L., 'Measurement and Enhancement of Waveform Digitizer Performance', *1976 Electro Conference Record*, New York.
5. Bernina, D., and Barger, J. R., 'High-speed, High-resolution A/D Converters: Here's How', *EDN* **18**, 62–6, June 5 (1973).
6. Tietze, U., and Nurmberger, H., Private communication. Lehrstuhl fur Technische Elektronik, Universität Erlangen-Nurnberg (1975).
7. Fiedler, H. L., Hoefflinger, B., Demmer, W., and Draheim, P., 'A 5-bit Building Block for 20 MHz A/D Converters', *IEEE Journal* **SC-16**, 151–5, June (1981).
8. Dingwall, A. G. F., 'Monolithic Expandable 6 bit MHz CMOS/SOS A/D Converter', *IEEE Journal* **SC-14**, 926–32, December (1979).
9. Emmert, G., Navratil, E., Parzefall, F., and Rydval, P., 'A Versatile Bipolar Monolithic 6-bit A/D Converter for 100 MHz Sample Frequency' *IEEE Journal* **SC-15**, 1030–32, December (1980).
10. Peterson, J. G., 'A Monolithic Fully Parallel, 8b A/D Converter', *1979 ISSCC Digest of Technical Papers* 128–9.
11. Lammert, M., and Olsen, R. K., '1-μm Shrinks and Speeds up Flash Converter', *Electronics* **55**, 135–7, May 5 (1982).
12. Koscielniak, H., and Seitzer, D., 'Eine neue Struktur für schnelle Analog/Digital-Umsetzer', *Nachrichtentechnische Zeitschrift* **29**, 535–7, July (1976).
13. Best, R. E., 'The Conversion Speed of Sequential and Sequential–parallel A–D Converters', *Scientia Electrica* **17**, 109–32 (1971).
14. Rollenhagen, D. C., 'High Speed A/D Converters for Radar Applications', *IEEE Proceedings ISCAS 1976*, 579–82.
15. Fletcher, R. E., 'A Video Analogue to Digital Converter', *Proceedings of the International Broadcasting Convention*, London, 1974, 47–57.
16. Crosby, P. S., 'The Design of a Highly Integrated 8-bit, 20 MHz Analog-to-Digital Converter for Automatic Measurements', *1977 Electro Conference Record*, New York, 37/5/1–5.
17. Schindler, H. R., 'Using the Latest Semiconductor Circuits in a UHF Digital Converter', *Electronics* **36**, 37–40, August 30 (1963).
18. Saul, P. H., 'Successive Approximation Analog-to-Digital Conversion at Video Rates', *IEEE Journal* **SC-16**, 147–51, June (1981).
19. Ninomiya, Y., 'An A/D Converter with 10 Bits/sample by 20 MHz Sampling Rate', *IEEE Trans.* **COM-28**, 1–6 January (1980).
20. Ninomiya, Y., 'An Accurate A/D Converter Sampling at 100 MHz', *IEEE Trans.* **COM-29**, 1353–7, September (1981).
21. Pratt, W. J., 'High Linearity and Video Speed Come Together in A–D Converters,' *Electronics* **52**, 167–70, October 9 (1980).
22. Graves, E. L., '5 MHz A–D Unit Converts 12 Bits', *Electronics* **51**, 185–6, August 30 (1979).
23. Smith, B. D., 'An Unusual Electronic Analog-Digital Conversion Method', *IRE Trans.* **PGI-5**, 155–60 (1956).
24. Waller, L., 'A–D Converter Hits 8 Gigahertz with GaAs Device', *Electronics* **51**, 43–4, March 29 (1979).
25. Hornak, Th., and Corcoran, J. J., 'A High Precision Component-tolerant A/D Converter,' *IEEE Journal* **SC-10**, 386–91, December (1975).
26. Bell Telephone Monograph 5057: Experimental 224 Mbit/s PCM-system. Murray Hill, NJ, USA: Bell Telephone Lab. (1965).
27. Zimmer, M., 'Genaue Analog/Digital-Umsetzer für die Hochfrequenz-Datenerfassung', *Elektronik* **28**, 41–5, June (1979).
28. Krull, K., 'Ein PCM-Codierer für Breitbandsignale', *Nachrichtentechnische Fachberichte* **42**, 92–103 (1972).
29. Woods, J. V., and Zobel, R. N., 'Fast Synthesised Cyclic-Parallel Analogue-digital Converter' *IEEE Proc.* **127**, Pt G., 45–51, April (1980).

30. Fiedler, U., and Seitzer, D., 'A High-speed 8 bit A/D Converter Based on a Gray-Code Multiple Folding Circuit', *IEEE Journal* **SC-14**, 547–51, June (1979).
31. Arbel, A., and Kurz, R., 'Fast Analog-digital-converter', *IEEE Transactions* **NS-22**, 446–9, February (1975).
32. Van de Plaasche, R. J., and van de Grift, R. E. J., 'A High-speed 7 Bit A/D Converter', *IEEE Journal* **SC-14**, 938–43, December (1979).
33. '10-Bit 20 MHz Video A/D Converter', *Analog Dialogue* **14-2**, 8 (1980).
34. Black, W. C., Jr., and Hodges, D. A., 'Time Interleaved Converter Arrays', *IEEE Journal* **SC-15**, 1022–9, December (1980).
35. Blandowski, R., Grau, J., and Uhlenbrock, W., 'Ein schneller 10-bit-Analog-Digital-Umsetzer mit MSI-Elementen', *Internationale Elektronische Rundschau* **23**, 57–63, March (1969).
36. McCreary, J. L., and Gray, P. R., 'All-MOS Charge Redistribution Analog-to-Digital Conversion Techniques—Part I', *IEEE Journal* **SC-10**, 371–9, December (1975).
37. Schmid, H., *Electronic Analog/digital Conversion*, Van Nostrand Reinhold, New Yor, (1970), Chapter 8.3.
38. Jenkins, J. A., 'The First Truly Monolithic 10 Bit Analogue to Digital Converter—Its Advantages and Application', *Microelectronics and Reliability* **16**, 389–94 (1977).
39. Saul, P. H., and Jenkins, J. A., 'A 10-bit Monolithic Tracking A/D Converter', *1978 ISSCC, Digest of Technical Papers* 138–9, 271.
40. Brokaw, A. P., 'A Monolithic 10-bit A/D using I^2L and LWT Thin-film Resistors', *IEEE Journal* **SC-13**, 736–45, December (1978).
41. Yamada, H., *et al.*, 'A 13 Bit Integrated Circuit ADC', *1979 ISSCC, Digest of Technical Papers* 180–81, 294.
42. Post, H. U., and Waldschmidt, K., 'A High-speed NMOS A/D Converter with a Current Source Array', *IEEE Journal* **SC-15**, 295–300, June (1980).
43. Boyacigiller, Z. G., Weir, B., and Bradshaw, P. D., 'An Error-correcting 12 Bit/20 μs CMOS A/D Converter', *1981 ISSCC, Digest of Technical Papers* 62–3.
44. Candy, J. C., 'A Use of Limit Cycle Oscillations to Obtain Robust Analog-to-Digital Converters', *IEEE Trans.* **COM-22**, 298–305, March (1974).
45. Candy, J. C., Ninke, W. H., and Wooley, B. A., 'A Per-channel A/D Converter having 15-segment μ-255 Companding', *IEEE Trans.* **Com-24**, 33–42, January (1976).
46. Ritchie, G. R., Candy, J. C., and Ninke, W. H., 'Interpolative Digital-to-Analog Converters', *IEEE Trans.* **COM-22**, 1797–1806, November (1974).
47. Dietrich, M., 'Zur Verbesserung des Signal-Rausch-Verhaltnisses von interpolativen Analog-Digital-Umsetzern', *NTG-Fachberichte* **65**, VDE Berlin (1978), 66–72.
48. Netravali, A. N., 'Optimum Digital Filters for Interpolative A/D Converters', *Bell Syst. Techn. J.* **56**, 1629–41 (1977).
49. Claasen, Th., *et al.*, 'Signal Processing Method for Improving the Dynamic Range of A/D and D/A Converters', *IEEE Trans.* **ASSP-28**, 529–37, October (1980).
50. Tewksbury, S. K., and Hallock, R. W., 'Oversampled, Linear Predictive and Noise-shaping Coders of Order $n > 1$', *IEEE Trans.* **CAS-25**, 436–47, July (1978).
51. Grandbois, G., and Pickerell, T., 'Quantized Feedback Takes its Place in Analog-to-Digital Conversion', *Electronics* **49**, 103–7, October 13 (1977).
52. Van de Plaasche, R. J., and van der Grift, R. E. J., 'A Five-digit Analog-Digital Converter', *IEEE Journal* **SC-12**, 656–62, December (1977).
53. Landsburg, G. F., 'A Charge-balancing Monolithic A/D Converter', *IEEE Journal* **SC-12**, 662–73, December (1977).
54. McCreary, J. L., and Gray, P. R., 'All-MOS Charge Redistribution Analog-to-Digital Conversion Techniques—Part I, *IEEE Journal* **SC-10**, 379–85, December (1975).

55. Fotouhi, B., and Hodges, D. A., 'High-resolution A/D Conversion in MOS/LSI', *IEEE Journal* **SC-14**, 920–26, December (1979).
56. Cooperman, M., and Zahora, R. W., 'Charge Redistribution Codec', *IEEE Journal* **SC-16**, 155–63, June (1981).
57. Yano, K., et al., 'A Per-channel LSI Codec for PCM Communications', *IEEE Journal* **SC-14**, 7–13, February (1979).
58. Everard, J. D., 'A Single-channel PCM Codec', *IEEE Journal* **SC-14**, 25–37, February (1979).
59. Pfrenger, E., Picard, P., and von Sichart, F., 'A Companding D/A-converter for a Dual-channel PCM-Codec', *1978 ISSCC, Digest of Technical Papers*, 186–7.
60. Euler, K., Schlichte, M., and Pfrenger, E., 'A PCM Single-channel Codec in LSI Technology with a 13-segment Characteristic', *Proceedings of the International Zurich Seminar on Digital Communications* **B2**, 1–4 (1974).
61. Huggins, J., Hoff, M., and Warren, B., 'A Single Chip NMOS PCM CODEC for Voice', *1978 ISSCC, Digest of Technical Papers* 178–9.
62. Kelley, S., and Ulmer, D., 'A Single-chip CMOS PCM Codec', *IEEE Journal* **SC-14**, 54–9, February (1979).
63. Blauschild, R. A., Tucci, P. A., and Russel, H. T., 'A Single-chip I^2L PCM Codec', *IEEE Journal* **SC-14**, 59–64, February (1979).
64. Caves, J. T., et al., 'A PCM Voice Codec with On-chip Filters', *IEEE Journal* **SC-14**, 65–73, February (1979).
65. Landsburg, G. F., and Smarandoiu, G., 'A Two-chip CMOS CODEC', *1978 ISSCC, Digest of Technical Papers* 180–81.
66. Cecil, J., et al., 'A Two-chip PCM CODEC for Per-channel Applications', *1978 ISSCC, Digest of Technical Papers* 176–7.
67. Wooley, B. A., and Henry, J. L., 'An Integrated Per-channel PCM Encoder Based on Interpolation', *IEEE Journal* **SC-14**, 14/20, February (1979).
68. Wooley, B. A., et al., 'An Integrated Interpolative PCM Decoder', *IEEE Journal* **SC-14**, 20–25, February (1979).
69. Croisier, A., and Jacquart, C., 'A Single Channel PCM Coder', *Proceedings of the 1972 International Zurich Seminar on Integrated Systems for Speech, Video and Data Communications* F2, 1–4.
70. Cantarano, S., and Pallottino, G. V., 'Logarithmic Analog-to-digital Converters: A Survey', *IEEE Transactions* **IM-22**, 201–13, Septembr (1973).
71. Ernst, P., Hybride Analogperipherie für digitale Mikroregler. PhD Thesis, Universität Erlangen-Nurnberg (1978).
72. Ernst, P., 'Hybride Analogperipherie für digitale Mikroregler', *NTG-Fachberichte* **68**, VDE Berlin, 1979, 60–63.
73. Schindler, H. R., 'Delta Modulation', *IEEE Spectrum* **7**, 69–78, October (1970).
74. Jayant, N. S., 'Digital Coding of Speech Waveforms: PCM, DPCM, and DM Quantizers', *Proc. IEEE* **62**, 611–32, May (1974).
75. Wehrmann, W., *Einfuhrung in die stochastisch-ergodische Impulstechnik*, Oldenbourg, Vienna/Munich, (1973).
76. Hirsch, J. J., Contribution à l'étude des conversions numérique analogique et analogique numérique au moyen d'une représentation stochastique de l'information. PhD Thesis, Université de Grenoble (1970).
77. Corradetti, M., and Oliva, I., 'MOS A/D and D/A Converter Circuits Based on the Stochastic Principle: Reliability and Economy for Industrial Control and Data Processing Systems', *Digest of 5. Mikroelektronik-Kongress*, Munich 1972: Oldenbourg, Vienna/Munich (1973), 314–26.
78. Roberts, L. G., 'Picture Coding Using Pseudo-random Noise', *IRE Trans.* **IT-8**, 145–54 (1962).
79. 'Capture Single Shot Transients and Repetitive Waveforms with Confidence', *Product Note 02-5952-7626*, Hewlett Packard.

80. Fisher, R. E., 'A 1200-Megabit per Second Microwave-carrier Gray-code Analog-to-Digital Converter', *IEEE Transactions* **MTT-16**, 541–47 (1968).
81. Bosch, B. G., 'Gigabit Electronics—A Review', *Proceedings IEEE* **67**, 340–79, March (1979).
82. Howes, M. J., and Morgan, D. V. (eds), *Charge-coupled Devices and Systems*, Wiley, Chichester (1979).
83. Linnenbrink, T., *et al.*, 'A One Giga Sample per Second Transient Recorder', presented at the Workshop on High Speed A/D Conversion 1980, 11–12 February, Portland, Oregon.
84. Deyhimy, I., and Eden, R., 'The GaAs CCD: A Candidate for an Ultra-high Speed Transient Digitizer', presented at the Workshop on High Speed A/D Conversion 1980.
85. Borsuk, G., *et al.*, 'High Speed CCD for Time Translation, presented at the Workshop on High Speed A/D Conversion 1980.
86. Jekowski, J., Hutton, J., and Villa, F., 'The Stripline Sampler—An Ultra-high Speed Transient Digitizer', presented at the Workshop on High Speed A/D Conversion 1980.
87. Allan, R., 'The Inside News on Data Converters', *Electronics* **53** 101–12, July 17 (1980).
88. Hoff, M. E., Jr, and Townsend, M., 'An Analog Input/output Microprocessor for Signal Processing', *1979 ISSCC, Digest of Technical Papers* 220–21.
89. Osborne, A., *Introduction to Microcomputers*: Vol. I; *Basic Concepts*, A. Osborne, Berkeley, Cal. (1976).
90. Kraft, D. G., and Toy, W. N., *Mini/Microcomputer Hardware Design*, Prentice-Hall, New Jersey (1979).
91. Camp, R. C., Smay, T. A., and Triska, C. J., *Microprocessor Systems Design*, Matrix Publishers, Portland (1979).
92. Webb, R. W., Cooper, F. R., and Randlett, R. W., 'A 12 Bit A/D Converter', *1981 ISSCC, Digest of Technical Papers* 54–5, 258–9.
93. Evanczuk, S., 'A/D Converter Pushes Gigabits, *Electronics* **55**, 48–9, June 16 (1982).
94. Kayanuma, A., *et al.*, 'An Integrated 16 Bit A/D Converter for PCM Audio Systems', *1981 ISSCC, Digest of Technical Papers* 56–7.
95. Caldwell, J., 'Digital Scope Invades Analog Domain', *Electronics* **55**, 131–35, April 21 (1982).
96. Rife, D., 'High Accuracy with Standard IC's: An Elegant ADC's Forte', *EDN* **27**, 137–44, April 28 (1982).
97. McGlinchey, G., 'A Monolithic 12b/3μs ADC', *1982 ISSCC, Digest of Technical Papers* 80–81, 296–7.
98. Sekino, T., Takeda, M., and Koma, K., 'A Monolithic 8 Bit Two Step Parallel ADC without DAC and Subtractor Circuits', *1982 ISSCC, Digest of Technical Papers* 46–7, 290.
99. Timko, M. P., and Holloway, P. R., 'Circuit Techniques for Achieving High Speed–High Resolution A/D Conversion', *IEEE Journal* **SC-15**, 1040–51, December (1980).
100. Singh, S. P., Prabhakar, A., and Battacharyya, A. B., 'C-2C Ladder Voltage Dividers for Application in All-MOS A/D Converters', *Electronics Letters* **18**, 537–8, June (1982).
101. Williams, J., 'Single-slope A–D Converter Makes a Comeback with 20-bit Linearity', *Electronics* **53**, 151–5, November 6 (1980).

CHAPTER 4

Digital-to-Analog Converters

4.1 Introduction

Digital-to-analog converters (also called decoders) reconstruct analog signals from their digital code words. Since quantization is irreversible and information gets lost thereby, D/A conversion is not just the inverse process of A/D conversion. Upon applying the n-bit digital word at the input of a D/A converter, a single discrete analog value is obtained at its output. This single value replaces an infinite number of originally continuous values of the original signal, i.e. it represents the original signal plus some quantizing noise. The inherent simplicity of the D/A conversion process leads to relatively simple architectures, as we will see. By means of the feedback structure, described in section 3.4, it also leads to a very efficient and simple A/D converter. Many A/D converters described in Chapter 3, such as the successive approximation converter, the stochastic converter, the interpolative converter, and some of the charge redistribution techniques make direct use of this fact.

Although several circuit techniques for implementing D/A converters have been published, only a few methods, mainly parallel-type-converters, are in use today. Serial-type converters, on the other hand, are slower, since their analog output is produced only after receiving all the digits of the input word in sequential form. Therefore they have limited use.

To reconstruct an analog signal from its digital counterpart a configuration as in Fig. 4.1 is used. In the block labelled D/A the actual reconstruction of a multi-level signal is done, whereby a train of samples having various amplitudes is produced. These pulses are then interpolated in a low-pass filter (recovery filter) to produce a continuous signal. If ideal low-pass filtering is assumed (Fig. 4.2) then the resulting signal V_r will have the same frequency spectrum as the original signal except for the added quantization noise.

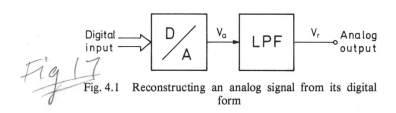

Fig. 4.1 Reconstructing an analog signal from its digital form

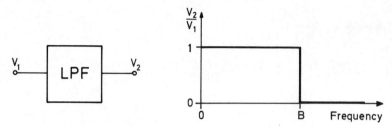

Fig. 4.2 Transfer function of an ideal low-pass filter

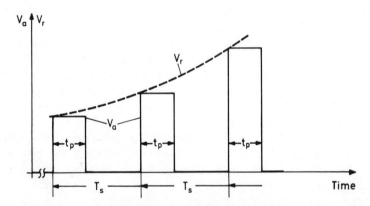

Fig. 4.3 Pulse train obtained at the output of a D/A converter

Actually, a D/A converter delivers a series of pulses having a certain width t_p and variable amplitudes V_a (Fig. 4.3). Moreover, in order to increase the energy content of the obtained signal, the pulse width t_p is usually expanded to cover the whole sampling period T_s. In this case, the output of the D/A converter may be viewed as a zero-order step interpolator (zero-order hold). The extension to a staircase function leads to a signal spectrum whose envelope is given by the function $y = \sin x/x$, where $x = \pi \cdot t_p/T_s$. Therefore, at the maximum frequency $B = 1/2T'_s k$ the signal exhibits a decay to $((\sin \pi/2)/\pi/2) = 0.637$. This deformation in the spectrum can be compensated by careful design of the interpolation filter. Detailed discussion of the approximation of a signal from a series of values using interpolation filters is given in reference 1.

4.2 Weighted Current Digital-to-Analog Converters

The weighted current D/A converter is also called a 'weighted resistor converter'. Its operation is based on the summation of binary weighted currents in a summing bus line connected to an operational amplifier circuit (Fig. 4.4). The resulting

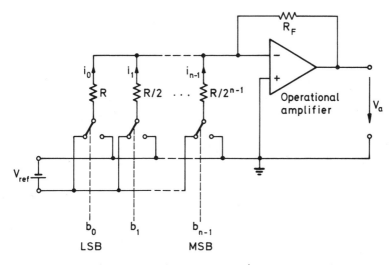

Fig. 4.4 A weighted current D/A converter

output voltage V_a is given by

$$V_a = -R_F \sum_{k=0}^{n-1} i_k$$

As shown in Fig. 4.4, the proper currents are produced through an array of n switches which are controlled by the n-bit digital input. The parallel input is applied to the switches such that the MSB-switch drives the maximum current when it is in the high state, i.e. '1', and, accordingly, the less significant bits are associated with lower weight currents.

Binary weighted currents can be obtained by applying a constant voltage V_{ref} to an array of weighted resistors having the values $R, R/2, R/4, \ldots, R/2^{n-1}$. A realization for the switches using bipolar transistors is shown in Fig. 4.5. A differential amplifier, operating as a current switch, drives the current i_k, according to the state of the proper bit, either to ground via T_1 or alternatively through T_2 to the operational amplifier circuit. (More information on switches is given in section 5.1).

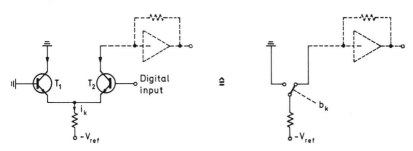

Fig. 4.5 Realization of a current switch and its equivalent circuit

The circuit features the following advantages. Independent of the digital input, the reference voltage source is continuously loaded by the same current. Because this constant current also flows during the switching time, no disturbing voltages due to transients occur in the reference line. In addition, since switching is effectively done between ground and the virtual ground of the operational amplifier, no change in the voltage drop across the reference resistor takes place. Hence, no parasitic capacitances are to be charged or discharged, giving rise to a short switching time.

The circuit of Fig. 4.4 suffers from the following disadvantages. The values of the required high-precision resistors vary considerably. This variation increases with increased resolution, so that a realization of the circuit in integrated circuit form becomes impractical, if not impossible.

The precision requirements on the resistor values also increase with increasing resolution. For example, the smallest resistor $R/2^{n-1}$ must be precise enough such that the maximum error in its current is smaller than the current i_0 corresponding to the LSB. To illustrate this fact, consider the case of a 10-bit converter. A tolerance of 0.05% corresponds to the LSB current. Due to temperature coefficients of resistors, the maximum power dissipation has to be limited in order to prevent the corresponding LSB current becoming comparable with the switches' leakage currents, thus giving rise to an additional source of error.

4.3 Ladder-Network Digital-o-Analog Converters

Another popular technique for D/A conversion uses the R–2R-resistive ladder network. As shown in Fig. 4.6, the circuit consists mainly of a resistor ladder having series resistors R and parallel resistors $2R$ and a bank of n equal-value switched current sources. The switching of the currents either to ground or to the respective node of the ladder is controlled, as before, by the digital inputs. The weighting action is provided here through the ladder by adjusting the attenuation from the sections of the ladder network. The ladder can be considered to be com-
to the LSB. To illustrate this fact, consider the case of a 10 bit converter. A

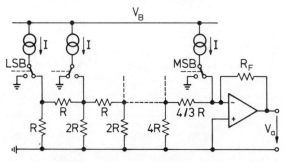

Fig. 4.6 Ladder network D/A converter (binary weighting)

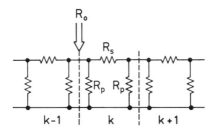

Fig. 4.7 Synthesis of the ladder network by a series of π-sections

resistor R_p are defined by the attenuation $a = i_k/i_{k-1}$ ($a < 1$) between adjacent sections and the impedance R_0 seen at each node:

$$R_s = R_0 \left(\frac{1}{a} - a\right) \quad \text{and} \quad R_p = 2R_0 \cdot \left[\frac{1 + (1/a)}{(1/a) - 1}\right]$$

For binary weighting $a = 1/2$, leading to $R_s = (3/2) \cdot R_0 = R$ and $R_p = 6 \cdot R_0 = 4 \cdot R$. Adjacent R_ps are combined to give 2R. So, the resistors' rate is $R_s : R_p = 1 : 2$.[2] The value of R depends on the manufacturing process and the operating speed wanted, because low impedance is needed to charge stray and load capacitance.

The terminating resistors on both sides should be selected so as to keep the condition of a constant input resistance R_0 at the corresponding nodes. Thus at the feeding point of the LSB current the proper terminating resistor is $R_T = 2R_0$. If combined with the already connected $R_p = 6R_0$, then the result is $3/2 R_0 = R$. Also, due to the virtual ground of the operational amplifier, the required values of the resistors R_p and R_T at the other end are $R_p = 6R_0$ and $R_T = 2R_0$, respectively. The value $R_T = 2R_0$ guarantees that the current I_{n-2} is attenuated by a factor of 2 before reaching the summing node of the operational amplifier, where the current I_{n-1} is fed directly, thereby maintaining the condition of binary weighting.

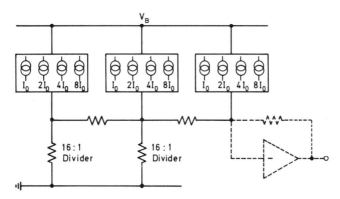

Fig. 4.8 Combining weighted-current and ladder network techniques to establish a 12-bit D/A converter

The drawback of this method is obviously the increased number of precision resistors, which is $(2n-2)$ as compared with n for the weighted resistor network. Therefore, practical converters of high resolution combine both techniques so as to have few resistors with manageable specifications, i.e. a smaller ratio of R_{max}/R_{min}.

Figure 4.8 illustrates the structure of a 12-bit D/A converter where both techniques are combined. Three identical current-source banks are required here, each composed of four binary-weighted current sources producing currents of I_0, $2I_0$, $4I_0$, and $8I_0$. By means of an attenuation of $1:16$ in the ladder, the current I_0 of the bank closer to the output on the right gives a factor of two more output currents than the largest current $8I_0$ of the bank adjacent to the left of it. If the attenuation is chosen to be $a = 1/10$, a D/A converter for the BCD-code (three decades in our example) is realized.[3,4]

Realization in integrated circuit form is feasible since, for example, the different currents could be obtained by paralleling a corresponding number of identical current sources producing the current of I_0 each. Identical transistor and resistor geometries are employed when designing the circuit layout. This ensures high precision and a constant current density all over the chip and, accordingly, good temperature stability. Errors due to temperature drift can be eliminated if high-β transistors (β > 300) and good matching of V_{BC} (within ±1 mV) are produced. In addition, current changes are compensated through a feedback operational amplifier circuit. The measured data for a 12-bit converter of this type at 25°C were a tempco = $5 \cdot 10^{-6}$ in the range $-55°C < T < +125°C$ and a settling time of 1.8 µs for a settling to 0.01% of full scale.

In a similar way, by combining current weighting for the three MSB's and R–2R attenuation for the 13 LSBs, a 16-bit converter is achieved.[20] In addition, laser trimming is applied for initial accuracy as well as bipolar dielectric isolation. Component tracking errors during temperature variations are minimized by equal bias conditions for all devices, feedback reference stabilization, and a bit cell layout that minimizes thermal gradients. A buried zener diode serves as a reference. Settling time for full-scale current to 0.003% is 1 µs.

In another design charge coupled devices (CCD) are proposed.[4] Thus instead of the current sources, a CCD shift-register is needed where the input data in form of 1's and 0's are stored. The taps of the CCD-register are connected to the nodes of the ladder network, thereby replacing the reference voltage. The summation is performed, after storing the input, by a summing amplifier and a sample-and-hold circuit.

In view of the progress of MOS technology with its high-quality capacitor capability, proposals have been made to use capacitive rather than resistive voltage dividers.[18] Further improvements to reduce the number of capacitors lead to a C–2C structure.[22]

A modification of the ladder network structure saves all the current sources and uses one voltage source instead, as shown in Fig. 4.9. Again, in the case of binary weighting a R–2R ladder network is used with a voltage source connected to the node closest to the end. The current $I = V_{ref}/R$ that is fed into the node with the

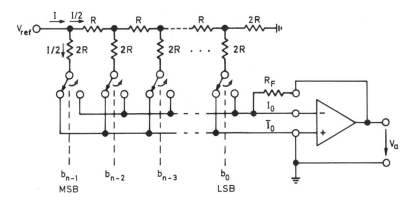

Fig. 4.9 Inverted R–2R ladder network D/A converter

impedance R, seen by the reference voltage, separates into two equal parts of value $I/2$ due to the impedances of 2R that are seen in parallel at that node. The part of $I/2$ flowing vertically is collected by the summing node of an operational amplifier. The part of $I/2$ flowing to the right meets twice 2R in parallel at the next node, and therefore separates into two equal parts, one of them $1/2 \times I/2$ flowing down again to the summing node. In other words, it is weighted by a factor 1/2 as compared with that flowing down one node to the left of it. The switches are, as explained earlier, controlled by the information bits to be decoded. V_{ref} meets a constant load R independent of information because the switches direct the currents either to ground or virtual ground, which is the amplifier summing node. For some applications it is useful to have access to the alternative summing line carrying $\bar{I}_0$, where

$$\bar{I}_0 = I_{max} - I_0$$

$$I_{max} = \frac{V_{ref}}{R}\left(1 - \frac{1}{2^n}\right)$$

The feedback resistor R_F is usually included on the ladder network chip for reasons of accuracy, i.e. temperature and long-term stability.

One of the problems associated with all of the binary weighted switched current converters is differential non-linearity (unequal step size) or non-monotonicity. This is again due to the fact that the currents associated with individual bits are more or less independent of each other. For example, when switching in a 4-bit converter from 0111 to 1000 (major carry transition) it is not easy to guarantee that the sum of all three less significant bits is just one LSB smaller than the MSB-weight current alone. This problem can be tackled in several ways. For instance, instead of adding together binary weighted values, one adds together equal amounts of current.[5] This means more complexity, because n current sources and switches are replaced by $(2^n - 1)$ combinations. This technique is, therefore, restricted to the two to four MSBs of high-resolution converters of more than 14 or 16 bits.[6,7] It is often referred to as a 'decoded' D/A converter.

In addition to the static error, a dynamic error might occur if not all of the switches open or close at the same time. Large transients can be seen on an oscilloscope display, known to and feared by all designers and users under the name of 'glitches'.

Current switching type D/A converters can achieve very high speeds. A 10-bit set-up with a settling time of 15 ns to $Q/2$ ($\hat{=} 1/2$ LSB) has been reported.[8] For still higher resolution of 12 bits, 35 ns are obtained.[9] For even higher resolution devices of 18-bit with 16-bit relative accuracy, the settling time increases to 2 μs.[6] The figures for the settling time relate to the output current into a low-impedance sink. For a finite impedance with additional capacitance loading, the settling time can increase considerably.

Maximum resolution by the end of 1980 was around 18 bits. In view of the rapid development of the state of the art and the great variety of suppliers and devices available, no other specifications are given here.

4.4 Voltage Divider Digital-to-Analog Converters

The problem of non-monotonicity described for the independent current switch converter can be overcome by using a series resistor chain as a reference voltage network. By this means a monotonic behaviour is guaranteed. This voltage divider is combined with a matrix of switches that connect the appropriate tap of the chain to the output (Fig. 4.10). Such a circuit has been implemented on a 300 mil² die area containing 256 resistors and 510 PMOS switches for an 8-bit D/A converter.[10] A 14-bit/40 μs D/A converter using a 7-bit coarse divider and a 7-bit fine divider in cascade is reported.[23]

4.5 Non-linear Digital-to-Analog Converters

Non-linear D/A converters are employed in floating point A/D converters, the majority of them being used for companding in PCM codecs for speech transmission (see sections 3.6.2 and 3.6.3).

The voltage divider D/A converter described in section 4.4 can be easily modified to provide a non-linear characteristic by suitable design of the resistor chain. By combining a non-linear main resistor chain with an auxiliary second linear divider which can be connected again by a matrix of switches a piecewise non-linear characteristic with linear subdivision can be implemented. An example of this type of converter is shown in Fig. 4.11.[11] Non-linear D/A converters can also be synthesized from either weighted resistor or ladder network decoders that basically deliver a current into their load. By suitable addition of currents into a summing node, the non-linear characteristic is obtained.

115

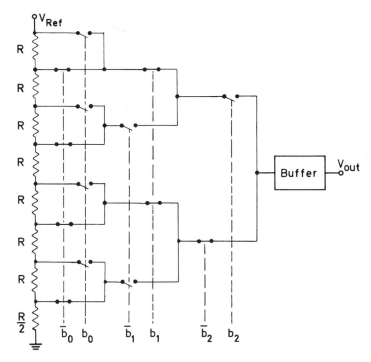

Fig. 4.10   Voltage divider D/A converter (3 bits)

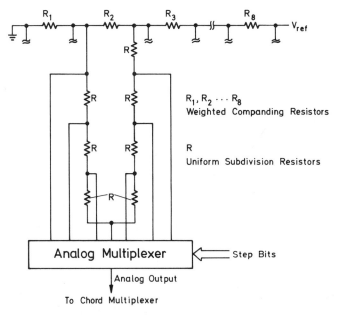

Fig. 4.11   Non-linear D/A converter

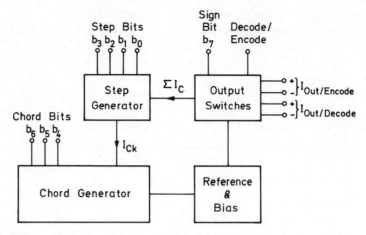

Fig. 4.12 Synthesis of the expanding characteristic (A-law, compare Fig. 3.39)

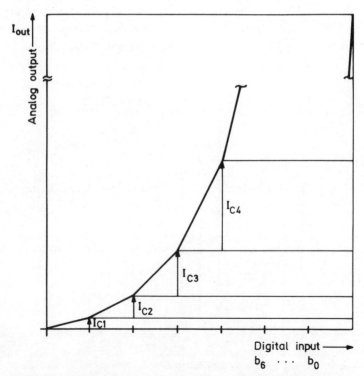

Fig. 4.13 Block diagram of a non-linear D/A converter based on current summation techniques

An example of this will be given, taking a decoder for PCM companding either according to the μ-law or A-law (see section 3.6.2). Figure 4.12 shows the output characteristic consisting of piecewise linear segments (= chords) that increase their steps by a factor of 2 by increasing the segment height accordingly. For simplicity, the sixteen steps within each chord are not shown. Starting from chord 1, the corresponding chord current I_{C1} is used as a pedestal upon which to build the next chord. In this way, monotonicity is obtained. Further subdivision within each chord into sixteen steps is achieved by a step generator that splits each of the currents I_{Ck} into thirty-three equal parts (thirty-three are needed instead of sixteen to properly establish consecutive pedestals, including half-step offsets when using the decoder in a feedback encoder.) This can be done accurately in an integrated circuit by choosing the emitter areas in proportion to the desired proportion of the current handled by the switch.

Figure 4.13 shows the basic circuit in a simplified block diagram. The chord current generator uses a R–2R ladder network for the four larger pedestals and a weighted current circuit for the smaller pedestal currents. The step current generator splits any value of pedestal current into the appropriate proportions that can be switched individually or as a whole. The output switches, together with the reference and bias block, provides bipolar and decode/encode operation of the whole circuit. More details can be found in references 12 and 13. A more generalized approach to the synthesis of non-linear D/A converters using linear decoders and switches as building blocks is given in reference 14.

4.6 Multiplying Digital-to-Analog Converters

The inverted ladder network D/A converter of Fig. 4.9 delivers an output voltage

$$V_{out} = -R_R \cdot I \cdot I \sum_{k=0}^{n-1} (b_k \cdot 2^k)$$

i.e. the output voltage builds up by summing all currents flowing into the bus line and through R_F. These are controlled by the digital inputs determining the coefficients b_k. As explained in section 4.3, the current I representing the LSB is

$$I = \frac{V_{ref}}{R \cdot 2^n}$$

Therefore

$$V_{out} = -V_{ref}/2^n \cdot R_F/R \cdot R_F/R \sum_{k=0}^{n-1} (b_k \cdot 2^k)$$

In section 4.3 it was assumed that V_{ref} is a constant voltage. However, if V_{ref} is variable (for instance, an a.c. voltage) then V_{out} is the product of V_{ref} times a factor that can be controlled by digital means, for example, by a microcomputer. This is the basis of a wide number of applications in process control or as an audio attenuator in hi-fi systems.[15]

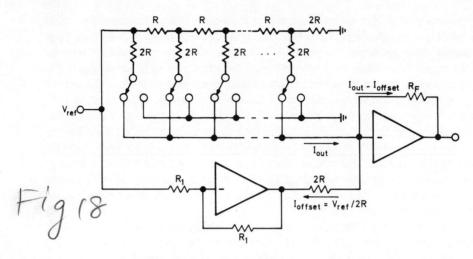

Fig. 4.14 Multiplying D/A converter structure for four-quadrant operation

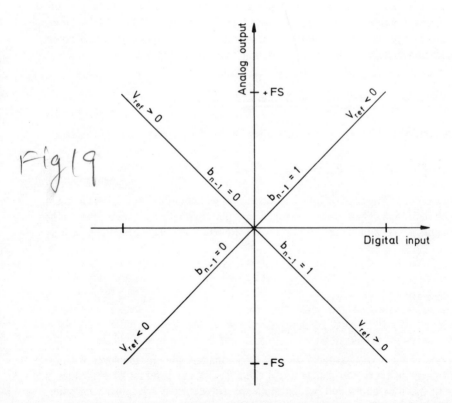

Fig. 4.15 Four-quadrant operation of the multiplying D/A converter of Fig. 4.14

The converter becomes a two-quadrant device with reversal of reference polarity. The device can now be operated as a four-quadrant multiplier by using an offset binary or sign plus magnitude code. For this use an offset current related to the reference voltage is introduced (Fig. 4.14). In this way the analog output extends from full-scale negative to full-scale positive between all ONEs and all ZEROs with operation in the first and third quadrant. Reversing the polarity of the reference changes the operation to the second and fourth quadrants, as indicated in Fig. 4.15.

Another useful modification is obtained if the weighted resistor network of a D/A converter is introduced into the feedback path of an operational amplifier instead of the constant resistor R_f of Fig. 4.9. In this case, the gain of the amplifier can be controlled in steps of a factor of 2 by digital inputs. This is useful for floating point A/D converters, as described in section 3.6.3.

4.7 Error Correction in Digital-to-Analog Converters

A major factor determining the accuracy of a D/A converter is the resistor variations that define the currents in either a weighted current or a ladder network converter. These have to meet stringent requirements, and therefore are usually made by either thick film or thin film resistor technology, where trimming is applied after completing the processing steps (for instance, by lasers that cut out a fraction of the cross-section or cut prefabricated links to correction resistors). A common way of tackling the non-monotonicity problem is to use $(2^n - 1)$ weights instead of n, as described in section 4.3.

Another method of correction that can be carried out even later in the production sequence (for instance, after packaging and sealing the whole device) is used especially in high-resolution devices of 14 bits and more. This is based on measuring the errors of the converter's input/output characteristic using the erroneous digital values as addresses for a correction PROM, i.e. a memory that delivers correction values to a correction D/A converter. From the point of view of accuracy, the most significant bit resistors are most critical with respect to relative accuracy. Therefore it is sufficient to take a correction DAC with a shorter word length. As an example,[16] a 12-bit correction DAC in connection with the five most significant bits has been used, i.e. the correction memory has a capacity of $(2^5 - 1) \times 12 = 372$ bits.

The same approach can be taken with respect to the correction of the gain error of the converter. This error is due to deviation of the feedback resistor R_F of the output buffer amplifier (Fig. 4.9) from its precise value. A gain correction 6-bit D/A converter is used, the resistor of which can be connected in parallel to the feedback resistor R_F under PROM control. In this way, the gain error can be reduced to below 1 LSB.

Depending on the type of PROM (either reversible or irreversible), the corrections can be repeated during the device's lifetime or are made only once.

If even higher resolution of 16 bits is required, a constant correction is not sufficient to guarantee the device specifications over a wider temperature range. In this case, a microprocessor-controlled calibration circuit that uses lower resolution is employed[17] to make differential linearity measurements. The calibration cycle can be initiated externally, and takes 3 s. During calibration, the converter is isolated from external signals and the microcomputer takes measurements of the converter's offset, the differential linearity, and gain. The readings are used to store correction codes in a 16-byte read/write RAM. Three additional D/A converters are introduced to adjust the gain, for trimming the offset and for calibration.

Another source of error in high-resolution converters is ground noise. If it depends on the coded information this can be cancelled by means of a ground current cancellation scheme that routes current under the control of an auxiliary R–2R ladder network from the supply voltage to the internal ground node.[7]

4.8 The Shannon–Rack Decoder

The Shannon–Rack decoder will be described as an example of a simple and ingenious solution of the D/A conversion process which is based on the periodic charging and discharging of a capacitor.

A simplified circuit diagram of this converter is given in Fig. 4.16. The constant current source I charges capacitor C through the switch S_1 which is controlled by the information bits. When closing switch S_2, discharging of C through resistor R starts. After a series of charging and discharging of C, controlled by the serial digital input, the remaining charge on C will represent the analog output.

A clock is needed to synchronize the operation of the decoder in conjunction with the binary input. As stated above, this input should be delivered in serial form starting with the LSB. Switch S_2 is periodically opened and then closed each clock period T. On the other hand, S_1 closes only during the first half-period $T/2$ of the clock to charge capacitor C if a logical '1' is received; otherwise it remains opened. In the second half-period it is always opened. Figure 4.17 illustrates the different states of the switches and the corresponding capacitor voltage for a digital input of 1001.

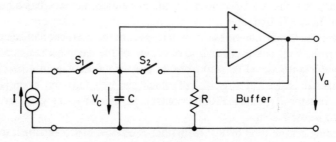

Fig. 4.16 The Shannon–Rack decoder

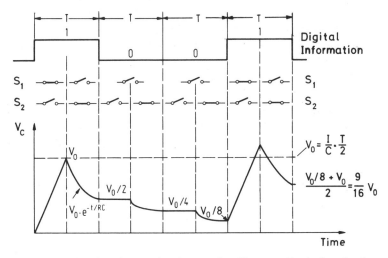

Fig. 4.17 Signal and control voltages of a Shannon–Rack decoder for a input sequence 1001

The required weighting by a factor 1/2 is provided here in each interval of length T if the discharge time-constant satisfies the condition

$$0.7 \cdot RC = T/2$$

which ensures that the capacitor voltage V_C is halved each half-period through discharging, as shown in Fig. 4.17. The analog output is provided by a voltage follower.

However, due to practical problems such as timing and the need of high-precision and low-drift components as well as serial digital input, such a converter has never been widely used.

4.9 Charge Redistribution Digital-to-Analog Converters

A modern version of the concept of charge weighting used in the Shannon–Rack decoder which is tailored for monolithic integration and is described in reference 18 under the concept of a charge redistribution decoder. In this method, the critical exponential decay of capacitor voltage to exactly one half of its initial value, depending on the timing and precision of the components, is replaced by connecting two identical capacitors in parallel while maintaining the charge on them constant.

Considering Fig. 4.16 again, if resistor R is replaced by another equal capacitor C, upon closing S_2, the stored charge is shared equally between both capacitors. In this case, Fig. 4.17 remains valid for the sequence but not for the capacitor voltage decay. The capacitor replacing resistor R has to be shunted by an additional switch for discharging the capacitor to ground during the time in which switch 1 is closed and switch 2 open. Capacitors can be well matched by modern integrated circuit techniques, such as MOS-technology, leading to proposals for using them as capacitive voltage dividers within converters.[22]

A D/A converter based on this modification was used in an integrated feedback A/D converter (Fig. 3.23). Using MOS-technology, the resulting chip area was 3 mm^2. A conversion time of 100 μs was measured for an 8-bit resolution. A drawback with respect to speed is the serial operation which needs $n \cdot (n + 1) \cdot T$ of conversion time T due to the algorithm.[18]

The same method of charging and discharging of capacitors under the control of the input data in a sequential manner (Fig. 4.17) could be carried out by using a cascade of sample-and-hold circuits under suitable timing and control.[19]

Another approach for a 14-bit DAC is based on the dual slope technique, described in section 3.5.2. Two currents, one for the seven MSBs and another for the seven LSBs, discharge a capacitor from full scale during a time that is controlled by a counter. The two currents are scaled with respect to each other by a ladder network. The bias can be trimmed by zapping of Zener diodes. Fabricated on two chips in a medium-speed bipolar process, a conversion time of 20 μs at 4 V full scale is achieved at a clock frequency of 12 MHz, the conversion time including a margin for resetting, initialization, and sample-and-hold acquisition at the output.[21]

4.10 A Note on Stochastic and Interpolative Digital-to-Analog Converters

Stochastic and interpolative D/A converters are closely related to the corresponding process of A/D conversion. Therefore they are treated in sections 3.6.6 and 3.4.4, in connection with A/D converters, respectively.

References

1. Garrett, P. H., *Analog I/O Design—Acquisition: Conversion: Recovery*, Reston Publishing Company, Reston (1981), Chapter 9.
2. Hoeschele, D. F., *Analog-to-Digital/Digital-to-Analog Conversion Techniques*, Wiley, New York (1978), Chapter 5.
3. Tatro, R. D., 'Which Hybrid Converter: Single-switch or Quad?' *Electronics* **47**, 89–93, July 25 (1974).
4. Weste, N., Mavor, J., and Madenman, D. J., 'Charge-coupled-device Digital-analogue Converter', *Electronic Letters* **11**, 551–2, October (1975).
5. Wilensky, S., 'Decoding Scheme Smooths 18-bit Converter's Nonlinearity', *Electronics* **53**, 128–32, June 5 (1980).
6. '18-bit Resolution with 16-bit Linearity Results from MDAC's Switching Scheme', *EDN* **26**, 73, January 21 (1981).
7. Guy, Th. S., and Trythall, L. M., 'A 16 Bit Monolithic Bipolar DAC', *1982 ISSCC Digest of Technical Papers*, 88–9.

8. Muto, A. S., et al., 'Designing a Ten-Bit, Twenty-Megasample-per-Second Analog-to-Digital Converter System', *Hewlett-Packard Journal* **33**, No. 11, 9–20, November (1982).
9. '12-bit Hybrid D–A Converter Uses Monolithic Design', *Electronics* **54**, 246, September 22 (1981).
10. Hamade, A. R., and Campbell, E., 'A Single Chip 8-bit A/D Converter', *1976 ISSCC Digest of Technical Papers* 154–5.
11. Hoff, M. E., Huggins, J., and Warren, B. M., 'An NMOS Telephone Codec for Transmission and Switching Applications', *IEEE Journal* **SC-14**, 47–53, February (1979).
12. Schoeff, J. A., 'A Monolithic Companding D/A-converter', *1977 ISSCC Digest of Technical Papers* 58–9.
13. Milojkovic, D., *Data Conversion with Companding DAC devices: Application Note*, Advanced Micro Devices, February (1978).
14. Shivaram, H. S., and Shivaprasad, A. P., 'A Simple Non-linear D/A Converter', *International Journal of Electronics* **51**, 63–9 (1981).
15. Hynes, M. J., and Burton, D. P., 'A CMOS Digitally Controlled Audio Attenuator for Hi-Fi Systems', *IEEE Journal* **SC-16**, 15–20, February (1981).
16. Brubaker, J., Boyacigiller, Z., and Bradshaw, P., '14-bit DAC Mates with µPs, Settles in less than 1 µs', *Electronic Design* **29**, 147–51, April 16 (1981).
17. Baresford, R., 'A Self-calibrating D–A Converter', *Electronics* **54**, 144, September 22 (1981).
18. Suárez, R. E., Gray, P. R., and Hodges, D. A., 'All-MOS Charge Redistribution Analog-to-digital Conversion Techniques—Part II', *IEEE Journal* **SC-10**, 379–85, December (1975).
19. Schmid, H., *Electronic Analog/digital Conversions*, Van Nostrand Reinhold, New York (1970), Chapter 7.3.
20. Guy, Th. S., and Trythall, L. M., 'A 16 Bit Monolithic Bipolar DAC', *1982 ISSCC Digest of Technical Papers* 88–9.
21. Mack, B., Horowitz, M., and Blauschild, R., 'A 14 Bit PCM DAC', *1982 ISSCC Digest of Technical Papers* 86–7, 300.
22. Singh, S. P., Prabhakar, A., and Bhattacharyya, A. B., 'C-2C Ladder Voltage Dividers for Application in All-MOS A/D Converters', *Electronics Letters* **18**, 537–8, June (1982).
23. Post, H.-U., and Schoppe, K., 'A 14 Bit Monontonic NMOS D/A Converter', *ESSCIRC '82 Digest of Technical Papers* 69–72.

CHAPTER 5

Devices and Building Blocks for Analog-to-Digital Converters

In our presentation of A/D conversion techniques the characteristics and operation of the various devices and basic circuits used in this field are considered here.

5.1 Switches

The voltage/current characteristic of an ideal switch is given in Fig. 5.1. A switch is said to be ideal if it renders:

(a) Zero impedance when it is on (closed), i.e. zero voltage drop independent of the current $I \lessgtr 0$;
(b) Infinite impedance in its off state (open) such that no leakage current flows independent of the voltage $V \lessgtr 0$ across the switch;
(c) No offset voltage, i.e. no voltage at zero current, or no offset current at zero voltage, in both ON and OFF states;
(d) Zero switching time in both directions, i.e. no delay;
(e) No interaction between the switched signal and the switch control;
(f) No loading to the controlling or driving source, respectively.

The traditional electromechanical switches are almost ideal unless speed is significant. In fact, it is impossible to find a switch having all these characteristics simultaneously. Therefore a wide variety of devices are used in practice, ranging from relays, semiconductor diodes, and transistors as well as thyristors, triacs, and thyratrons to mercury-arc rectifiers. The designer has to choose the proper

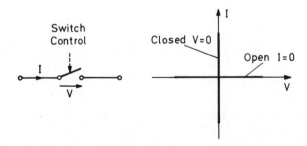

Fig. 5.1   Characteristic of an ideal switch

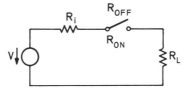

Fig. 5.2   Series switch

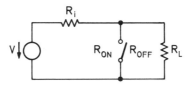

Fig. 5.3   Parallel switch

switch, depending upon the application. Here we will restrict ourselves to those which are commonly used in signal processing.

It is useful to keep in mind that the circuit environment of a switch can help to overcome or smooth its non-ideal properties. As an example, consider the series switch in Fig. 5.2. If the load resistance R_L is chosen such that $R_L \gg R_{ON}$, the ON-resistance of the switch, then the voltage drop across the switch and, accordingly, the attenuation, when it is closed, is small. Also, if $R_L \ll R_{OFF}$, the switch OFF-resistance, then the attenuation is increased. The situation is reversed in the case of a parallel switch (Fig. 5.3). Moreover, if it is not possible to obtain the required performance for the OFF and ON conditions by a single switch, then a combination of series and parallel switches can be used.

Switching applications can be classified into digital and analog switching. A digital switch always carries the same current in its ON-state whereas in an analog switch the current flowing depends on the magnitude of the signal. Therefore the linearity of their characteristics is of prime importance. Examples of digital switches are those used in D/A converters. Analog switches are used, among other applications, in sample-and-hold circuits. The well-known analog switches, used extensively in telephone exchanges (Reed relay, rotary and cross-bar switches) are still difficult to replace by all-electronic counterparts due to their high performance requirements. However, electronic switches are smaller in size, lighter, have a longer lifetime, are lower in cost, and have considerably higher switching speeds than mechanical switches and also have negligible jitter.

5.1.1   Mechanical Switches

Mechanical switches feature many properties of ideal switches. In its ON-state, the forward resistance of a mechanical switch lies in the range of milliohms with a zero offset voltage. The OFF-resistance and therefore the attenuation is almost infinite. Attenuation might decrease at high frequencies due to stray capacitances. Therefore switched signals can have frequencies of up to 10 GHz. In addition, since such switches are magnetically actuated, the interaction (feedthrough) between the switched signal and the switching control is negligible. However, a certain amount of control power has to be sacrificed. The main disadvantage of mechanical switches is their low switching speed, which lies in the range of milliseconds. In addition, the delay time between the ON-command and the ON-state varies considerably (jitter) in consecutive events so that it is difficult to specify. A major disadvantage, from the point of view of modern circuit realizations, is its relative large dimension compared with other electronic components surrounding them.

Therefore an electronic replacement for the mechanical switch which is competitive without too many drawbacks has been sought for a long time. This was successfully achieved only in cases where properties are required that could not be provided by a mechanical switch, e.g. a predetermined switching time. A well-defined and fast switching time is of great importance in A/D and D/A converters. Semicondutor devices are used mainly today in the implementation of switches for the various applications in A/D converters.

5.1.2 Semiconductor Diodes

The semiconductor diode is the simplest form of an electronic switch. As shown in Fig. 5.4, the characteristic of such a simple switch deviates appreciably, due to its exponential nature, from that of the ideal switch. Its main drawbacks are therefore the voltage drop (0.3–1 V) and the non-linearities in the forward direction. Also, its dynamic behaviour, as shown in Fig. 5.5, is completely different from ideal. This deviation from the static curve starts to become noticeable in the order of

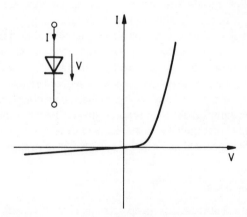

Fig. 5.4 Static characteristic of a semi-conductor diode

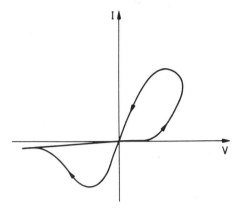

Fig. 5.5 Dynamic characteristic of a semi-conductor diode

several ten kilohertz. In addition, precautions have to be taken to prevent the effects of switch control from influencing the switched signal.

Metal-semiconductor junctions (e.g. Schottky or hot-carrier diodes) are most suitable due to their small ON-state voltage, high OFF-resistance, and negligible charge storage, i.e. high switching speed.

5.1.3 The Bipolar Transistor Inverted Mode Switch

First, a common emitter transistor circuit operating in its saturation region with a load resistance, R_L is considered. As shown in Fig. 5.6(b), in this case the voltage drop across the transistor at its operating point is V_{CEsat}. This drop depends on the battery voltage V_{CC}, the load resistance R_L, and the I_C-V_{CE} characteristics of the transistor. A simple equivalent circuit is given in Fig. 5.6(c), whereby the transistor is replaced by a switch symbol.

To study the behaviour of the transistor in the saturation region the model of

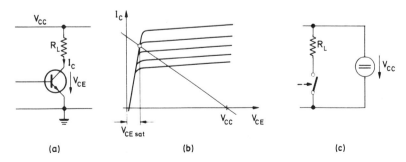

Fig. 5.6 A bipolar transistor switch. (a) Circuit; (b) operating point; (c) equivalent circuit

Ebers and Moll,[1] shown simplified in Fig. 5.7, is considered. Neglecting the series resistors, the following equations can be established:

$$I_E = -I_F + \alpha_R \cdot I_R$$
$$I_C = \alpha_F \cdot I_F - I_R$$

where α_F is the forward common-base current gain and α_R is the inverted common-base current gain (i.e. collector and emitter interchanged).

Applying the equations of an ideal p–n junction for the emitter–base and collector–base diodes we get:

$$I_F = I_{ES} (e^{-V_{EB}/V_T} - 1)$$
$$I_R = I_{CS} (e^{-V_{CB}/V_T} - 1)$$

in which I_{ES} and I_{CS} are the reverse saturation-currents for the emitter–base and collector–base junctions, respectively, and $V_T = 25.9$ mV at room temperature. Recalling that $I_B = -(I_C + I_E)$, from the formulae above we get

$$I_B = (1 - \alpha_F) \cdot I_F + (1 - \alpha_R) \cdot I_R$$
$$= I_{FB} + I_{FR}$$

Rearranging Fig. 5.7, as shown in Fig. 5.8, it can be concluded that one saturated transistor can be represented by two active transistors with the collector of one transistor connected to the emitter of the other, and vice versa, each of them being controlled by its associated base current shares I_{FB} (forward transistor) and I_{RB} (reverse transistor). Provided that the two transistors are operating in their active regions with the base currents I_{FB} and I_{RB}, the collector current is then

$$I_C = \frac{\alpha_F}{1 - \alpha_F} \cdot I_{FB} - \frac{1}{1 - \alpha_R} \cdot I_{RB}$$

Since

$$V_{CE} = V_{CB} - V_{EB}$$

it is possible to draw the relation $I_C = f(V_{CE})$ with I_B as a parameter in the satura-

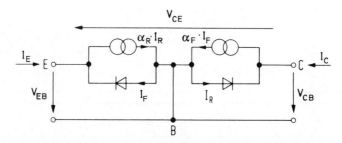

Fig. 5.7 Simplified model of a saturated transistor after Ebers and Moll[1]

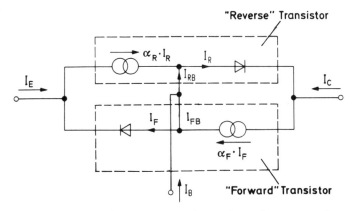

Fig. 5.8 Representing a saturated transistor by two active transistors

tion region, as shown in Fig. 5.9. The transistor circuit, including currents and voltages, is drawn in the same figure. Applying the reciprocity relation

$$\alpha_R I_{CS} = \alpha_F I_{ES}$$

which is valid in the saturation region,[2] then we get for the voltage drop V_{CE} across the switch

$$V_{CE} \approx V_T \ln \frac{1 + (I_C/I_B)(1 - \alpha_R)}{\alpha_R [1 - (I_C/I_B)(1 - \alpha_F)/\alpha_F]}$$

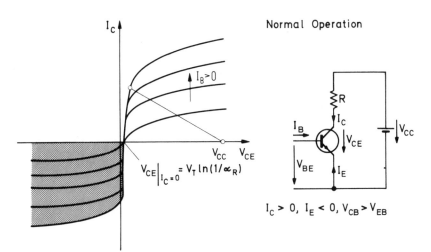

Fig. 5.9 Circuit diagram and characteristics for a transistor operating in the normal mode

130

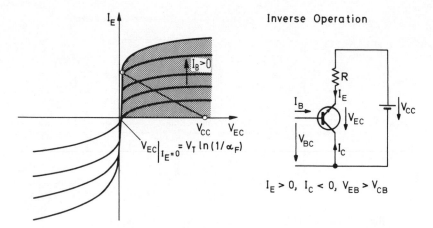

Fig. 5.10 Circuit diagram and characteristics for a transistor operating in the inverse mode

The offset voltage V_{CEO} is obtained by inserting $I_C = 0$ in the above equation, yielding

$$V_{CEO} = V_{CE}(I_C = 0) = V_T \ln \frac{1}{\alpha_R}$$

Since α_F is larger than α_R, it is to be concluded that if the transistor switch is operated in its inverted connection a very small offset voltage is obtained. For example, if the inverted mode current gain α_R of a certain transistor is 0.9, then the resulting voltage is $V_{CEO} = 2.74$ mV, while it is only 0.26 mV if the transistor is operated in its inverted mode with $\alpha_F = 0.99$. The obtained offset voltage is thus nearly one order of magnitude smaller.

This conclusion is further illustrated if the transistor characteristics, shown in Fig. 5.9, are extended beyond $I_C < 0$ for the range $V_{CE} < 0$. In the case of collector and emitter being interchanged, the relation $I_E = -(I_C + I_B)$ is needed to draw the characteristic curves $I_E = f(V_{EC})$, with I_B as a parameter for drawing the load line of the reverse operation circuit of Fig. 5.10. Since $I_B > 0$, having a value, near the axis intersection point, of the same order of magnitude as both I_E and I_C the whole set is shifted upwards for $I_C < 0$, thereby shifting the intersection point $V_{CE}(I_E = 0)$ towards the origin. The result is $V_{EC}(I_E = 0) < V_{CE}(I_C = 0)$, i.e. a smaller offset voltage than in the normal mode of operation.

In order to complete the analogy to the case of normal operation, the voltage V_{CE} is to be replaced by V_{EC} so that the third quadrant of $I_C = f(V_{CE})$ becomes the first quadrant of the relation $I_E = f(V_{EC})$. However, due to the small current gain α_R, the characteristics for different values of I_B = constant are closer to each other.

5.1.4 The Saturated Emitter Follower

By extending the transistor characteristics beyond the axis intersection point $I_E = 0$ it is easy to conclude that the voltage drop over the transistor (switch) for the range $I_E < 0$ can assume a zero value (Fig. 5.11). The circuit can be considered in this case, with its emitter resistor R, as an emitter follower operating in its saturation region. The relatively large slope of the characteristics in this region is advantageous, since the switched signal V_{CC} can vary considerably without any significant change of the zero voltage drop on the switch terminals (collector and emitter).

An electronic switch using complementary transistors operating in inverted mode is described in reference 3. The circuit was designed for an 11-bit D/A converter with bipolar output and a TTL-compatible input. A voltage drop of ± 10 mV was measured for current smaller than 6 mA. The difference in voltage between the *pnp*- and *npn*-sides is smaller than 1 mV. The measured temperature drift over the range 25°C–115°C varies between 8–22 µV/°C. The turn-on time with a 3 kΩ load is 10 ns. A nearly equal value for the turn-off time was measured. Also, the settling time for the resolution (11 bits) is 100 ns. The turn-on time was observed to vary between 50 and 65 ns only, for a capacitive load ranging between 15 pF and 680 pF. Fast D/A converters are now obtainable in hybrid or fully integrated form. The best types have settling times of about 35 ns for 12-bit,[4] 12 ns for 10-bit, or 5 ns for 8-bit.[5]

5.1.5 The Current Switch Circuit

A digital circuit switch that is widely used in high-speed non-saturated logic such as ECL and in D/A converters is shown in Fig. 5.12(a). This consists of a

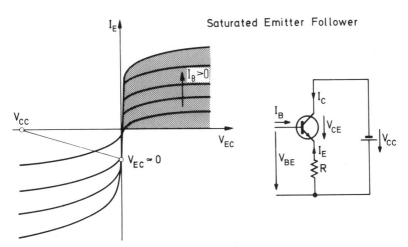

Fig. 5.11 Circuit diagram and characteristics of a saturated emitter follower

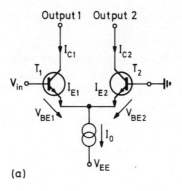

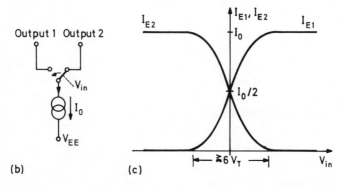

Fig. 5.12 Current switch. (a) Circuit; (b) equivalent circuit; (c) characteristic

differential amplifier pair of transistors T_1 and T_2 (also called a 'longtail pair'), the emitters of which are connected to each other and fed by a constant current source I_0. Therefore

$$I_{E1} + I_{E2} = I_0$$

Each emitter current is controlled by its base emitter voltage $V_{BE1,2}$, i.e.

$$I_{E1} = I_{S1}(e^{V_{BE1}/V_T} - 1), I_{E2} = I_{S2}(e^{V_{BE2}/V_T} - 1)$$

with $I_{S1,2}$ = reverse saturation current of emitter-base junction and $V_T = 25.9$ mV at room temperature.

After a brief calculation, taking into account that for a forward biased junction $I_{S1,2} \ll I_{E1,2}$, it follows

$$I_{E1} = \frac{I_0}{1 - e^{V_{in}/V_T}}, I_{E2} = \frac{I_0}{e^{V_{in}/V_T} - 1}$$

These relationships are plotted in Fig. 5.12(c). Due to its exponential nature, the transition region between the limiting values $I_{E1,2} = I_0$ and $I_{E1,2} = 0$, respectively, are only about 6 $V_T = 150$ mV. Therefore a small control voltage of slightly more than 150 mV is sufficient to switch I_0 from transistor T_1 to transistor T_2. As a switch, the circuit is operated in one of its limiting or extreme positions, steering a constant current I_0 into one of two outputs (Fig. 5.12(b)). This is the reason for its name.

As a differential amplifier, the circuit is operated near $I_{E1,2} = I_{0/2}$ as a linear amplifier. Common mode voltages are rejected as long as I_0 is an ideal current source, because such voltages change the voltage drop across I_0, leaving the transistor V_{BE}'s unchanged. A temperature change, equal for both transistors, acts as a common mode signal. Therefore the circuit is very stable with temperature and is used as an input stage for operational amplifiers and comparators.

As a current switch, when overdriven the non-conducting transistor is cut off and the conducting transistor does not get into saturation as long as the collector-to-base voltage is sufficiently large. This is easy to design because, with a collector resistor R_C, the voltage drop is limited to $R_C \cdot I_0$. Due to its non-saturated operation the circuit can switch very fast, in the order of a few nanoseconds.

5.1.6 Opto-electronic Switches

Combining a bipolar transistor and a semiconductor light emitting diode (LED) produces an efficient switch. The switch control signal is used here to drive a forward current through the LED. The emitted light, in turn, switches the transistor on, thereby connecting the switch terminals. This light-controlled switch (opto-isolator switch), shown in Fig. 5.13, has many advantageous properties such as low offset voltage, relatively high-speed switching (since it is current-driven), high isolation between the control and the switch circuit, and the fact that it can be produced in integrated circuit form.

5.1.7 The Field Effect Transistor as a Switch

Field effect transistors (Junction; MOS–FET, CMOS) show many desirable characteristics as electronic switches. They have a very low offset current, nearly

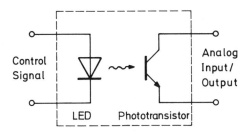

Fig. 5.13 Opto-electronic switch

zero offset voltage, good linearity, a controllable ON-resistance, and a very high input-impedance, so that a small driving power is sufficient.

Figure 5.14 illustrates the different types of FET-transistors together with their circuit symbols and transfer characteristics. According to the type of the substrate semiconductor, there are p-channel and n-channel Field Effect Transistors (FETs) which differ only in the polarities of the currents and electrode voltages. Important for the specific application is the property of being normally 'on' (depletion type) or normally 'off' (enhancement type). Normally 'on' means that at zero control voltage between gate and source, i.e. $V_{GS} = 0$, the channel between drain and source is conducting. All-junction FETs (JFET) are of the depletion type. Their channel resistance increases by increasing the reverse bias voltage V_{GS} (negatively for n-channel) until the drain current I_D reduces to zero at the 'pinch-off' voltage V_P. Reversing the bias polarity must be avoided, since in this case the input junction (gate–source) becomes forward biased, thereby destroying the high input impedance.

MOSFETs, on the other hand, feature high input impedance, for both polarities of V_{GS}. This is due to the good isolation provided by an oxide layer between the gate electrode and the channel.

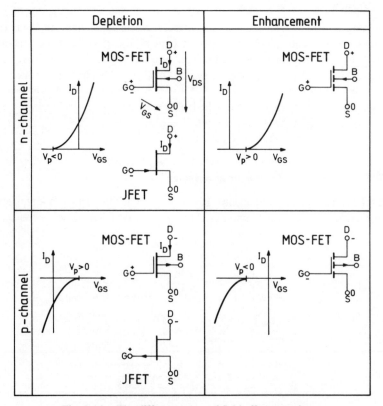

Fig. 5.14 The different types of field effect transistors

A simplified analysis of the current/voltage relationships leads to the following equation for the drain current in the so-called ohmic region, i.e.

$$0 \leq V_{DS} \leq V_{GS} - V_P \text{ (} n\text{-channel)}$$

or

$$0 \geq V_{DS} \geq V_{GS} - V_p \text{ (} p\text{-channel), respectively:}[6]$$
$$I_D = \beta [(V_{GS} - V_p) \cdot V_{DS} - V_{DS}^2/2]$$

in which

$$\beta = \mu \cdot C_{0X} \cdot W/L$$

and μ = the carrier mobility in the channel, C_{0X} = capacitance per unit area of the MOS-capacitor, W = channel width, perpendicular to current flow, and L = channel length, in the direction of current flow.

As an example, the characteristics of a n-channel enhancement mode transistor are shown in Fig. 5.15. The characteristics are parabolas, having their vertex at

$$V_{DS} = V_{GS} - U_p \text{ and } I_D = (\beta/2)(V_{GS} - V_P)^2.$$

For values of $V_{DS} > V_{GS} - V_P$, the drain current remains constant and the transistor is said to operate in its pinch-off region.

For small values of $V_{DS} \ll V_{GS} - V_P$, the drain current is given approximately by

$$I_D \approx \beta \cdot (V_{GS} - V_P) \cdot V_{DS}$$

This linear relationship, that explains the name 'ohmic region', is drawn in Fig. 5.16 for a p-channel JFET.[7] The characteristics are symmetrical with respect to and in the vicinity of the origin, and represent, in fact, a controllable ohmic resistance with V_{GS} as the controlling parameter. As an example, the channel resistance for the transistor at $V_{GS} \simeq 0.8 \cdot V_P$ reaches about ten times its value at $V_{GS} = 0$. This means that channel resistance varies considerably if the signal

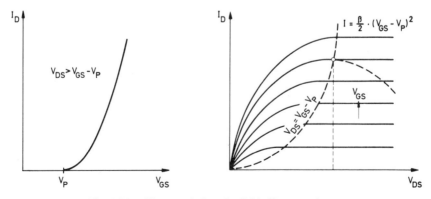

Fig. 5.15 Characteristics of a field effect transistor

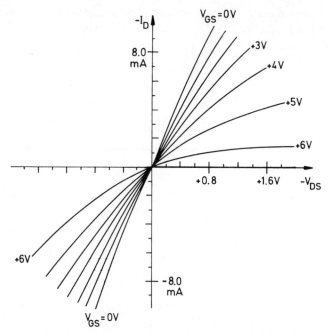

Fig. 5.16 Drain characteristics of a *p*-channel JFET in the vicinity of the origin

voltage amplitude is comparable with V_P. This can seriously affect accuracy. In this case it is advisable to use a switching or chopper FET which has a larger ohmic range. Since we are dealing with a semiconductor 'resistor', a temperature coefficient of about $7 \cdot 10^{-3}/\mathrm{K}$ is to be expected. Typical values of channel resistance lie in the range of 100Ω, although smaller values down to a few ohms are possible through proper selection of the transistor dimensions. Typical switching times for a single-transistor switch range from 10 ns to 100 ns. For circuits with an integrated drive circuit, switching times of 0.5–1 μs are common (Table 5.1).

Usually, a circuit as shown in Fig. 5.17(a) is used to drive a FET-switch. If the diode is off, the voltage drop across R and V_{GS} are both zero. The channel is correspondingly on and the switch is closed. The amplitude of V_C must be high enough to keep the diode off even at the postive peaks of the switched signal. To turn the switch off, V_C should be reduced to a value that can forward bias the diode, reversing V_{GS} to a value below switch-off. This value of V_C must be negative such that V_{GS} remains well below V_P for all values of the signal voltage.

As mentioned earlier, the channel ON-resistance primarily determines the switch series-resistance encountered by the signal source. Resistor R in Fig. 5.17(a) is introduced to speed up the discharging of the gate–source capacitance which starts after the diode has been turned off. The originally high input impedance of the switch in the off position is mainly determined by the resistor R.

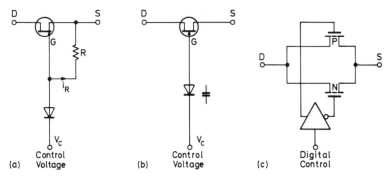

Fig. 5.17 Analog FET switches. (a) JFET switch; (b) VARAFET switch; (c) CMOS switch

p-channel JFETs feature no quiescent current and they are usually driven by TTL or CMOS logic. n-channel FETs, on the other hand, render a lower ON-resistance but they need an appreciable driving power, which is usually provided through a bipolar or a bipolar/MOS combination.

The use of an NJFET, by itself, as a switch is not satisfactory in some applications; for example, in sample-and-hold circuits. This is due to the fact that upon turning the switch on, some charges are injected into the signal path, leading to a distortion in the output. A varactor diode is therefore usually introduced in order to limit these injecting charges by storing them. The resulting device is called the VARAFET (Fig. 5.17(b)).[8]

Field effect transistors are used today in the manufacture of different types of switches in monolithic integrated circuit form.[9] This ranges from the single pole ON/OFF switch (*S*ingle *P*ole *S*ingle *T*hrow—SPST) and the DPDT (*D*ual *P*ole *D*ual *T*hrow) switches to the 16-pole multiplexer having a 4-bit address. MOSFET

Table 5.1 Comparing some types of analog switches[9]

		Input range (V)	ON-resistance (Ω)	Switching times t_{ON}/t_{OFF}	Steady state control current (mA)
Reed relays	±	300	0.1	1.0 ms/2.0 ms	10
Bipolar switches discrete elements	±	12	1–10	10 ns/50 ns	10
IC (hybrid) JFGT with TTL-drive	±	8	30	0.5 µs/10 µs	3
IC (hybrid) MOSFET with TTL-drive	±	10	100	0.5 µs/1 µs	3
Monolithic IC CMOS	±	14	75	1.0 µs/0.5 µs	0.1

switches are used only in some specific applications. Their single outstanding property is their large packing density, i.e. there are more switches per package (up to sixteen are available) when produced as monolithic ICs. The dominant types of switches are now the CMOS switches (Fig. 5.17(c)). Beside being available in monolithic form, they have an extremely low driving power, as shown in Table 5.1, where the various switch types are represented by typical values.

A D/A converter that uses JFET switches with a bipolar transistor drive is described in reference 7. With a load resistance of about 50 kΩ, the measured static error was less than 10^{-4} for each switch.

5.1.8 Analog Switches Using Diodes

A combination of an operational amplifier with semiconductor diode switching circuits can improve their characteristics as a switch, e.g. bandwidth, linearity, and ON- and OFF-characteristics, thereby giving high-quality switches.[10]

As shown in Fig. 5.18, the switch is closed if the two points A and B are connected. Assuming an ideal operational amplifier, the output is given by:

$$V_2 = -V_1 \cdot R_{F1}/R_1$$

Imperfections of the switch A–B–C no longer have any influence on the output, since it is mainly determined by the resistance ratio R_{F1}/R_1. Moreover, the ON-resistance can be compensated for by selecting the ratio R_{F1}/R_1.

The switch is opened if the two points A and C are interconnected. Since a feedback loop is still provided through R_{F2} in this case the input voltage V_1', due to the virtual ground, and therefore the output voltage V_2, are also negligibly small. Furthermore, in the case of finite amplification of the amplifier, the voltage V_2 can be made zero if a suitable resistor is connected across B and C to compensate for the small current through R_{F1}.

The switch A–B–C is established by a diode bridge (Fig. 5.19). With positive voltage V of the pulse generator, the upper bridge conducts, thereby connecting the two points A and B. The ON-resistance of the switch in this case is the small forward resistance of the diodes. This resistance is approximately constant and

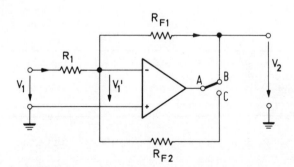

Fig. 5.18 Feedback gate with operational amplifier

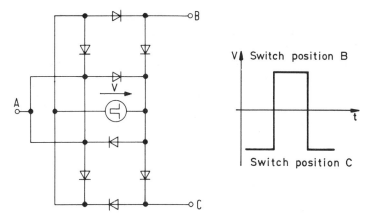

Fig. 5.19   Implementing the switch with diodes

independent of the value of the current, since any increase in the current in one branch of the bridge is accompanied by a corresponding decrease in the current of the other. The overall resistance therefore remains constant. If the polarity of V is changed, the lower bridge conducts so as to connect points A and C.

In an experimental set-up of such a circuit the measured attenuation in the ON-state was 0 dB while reaching 100 dB in the OFF-state. These values are maintained up to 10 MHz.[10]

5.2   Sample-and-Hold Circuits

5.2.1   Introduction

The function of a sample-and-hold circuit (S/H) is to take, according to the sampling theorem (section 2.1), samples from a continuous voltage and then to hold the level for a certain time interval. During this 'hold' period, further processing of the sampled waveform takes place, as A/D conversion for example.

In this section, useful definitions as well as the relationships describing the associated errors in a S/H process are presented. A useful criterion for the total error is that given by quantization. Low-loss capacitors are used as an inter-

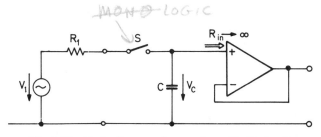

Fig. 5.20   Basic diagram of a sample-and-hold circuit

mediate storage medium. Accepted types are mainly made of polystyrene, polycarbonate, polypropylene, and Teflon. MOS-capacitors are also suitable, especially for hybrid S/H circuits.[11]

A simplified S/H circuit is shown in Fig. 5.20. The signal source, having an internal resistance R_1 and an open-circuit voltage V_1, charges capacitor C through the switch S. During the charging period t_A (called the acquisition time), a sample is taken. Afterwards the switch is turned off for the hold period t_H during which the sampled amplitude is kept constant. To avoid loading of the hold capacitor C by any output circuit, a buffer amplifier is used. If necessary, a second buffer stage is introduced at the input of the S/H circuit so that C is charged without loading the signal source. The sum of t_A and t_H represents the actual sampling period T_S which has to fulfill the condition $T_s < 1/2B$ due to the sampling theorem, where B is the bandwidth of the input signal.

In practice, the sampling process deviates from theory. For instance, the sampling pulse has a finite width t_A (Fig. 5.21(a)) during which the capacitor charges to the value of the signal voltage (within a certain error band) through the resistance $R = R_1 + R_S$, with R_S the switch ON-resistance.

If the time constant could be made such that $RC \to 0$, the S/H would act as an analog gate during the sampling interval t_A. This operation is called 'natural sampling',[12] and there is no attenuation of the signal, depending on its frequency as long as it is situated within the bandwidth B. Holding the sampled value during the time $0 < t_H < T_s$ (flat-top sampling) a shaping of the signal spectrum given by the function

$$\frac{\sin \pi f_x t_H}{\pi f_x t_H}$$

occurs as already given for the output signal of a D/A converter in Chapter 4 (f_x is the frequency of the input signal).

For $RC \ll t_A$ the capacitor voltage approaches steady state during the sampling interval. Assuming a nearly constant signal level within this time, then C charges exponentially so that the signal is reached within an accuracy of $-Q/2$. This condition leads to

$$e^{-t_A/RC} \leq 2^{-(n+1)}$$

if the S/H is used in front of an n-bit A/D converter. In the case of 8-bit, it is $t_A = 6.24 \cdot RC$.

The first-order low-pass response of the RC-section in the tracking mode (switch closed) must also be considered. For a $-Q/2$ decrease of the amplitude at the maximum signal frequency $f_x = B$, the 3-dB bandwidth has to be

$$f_3 = B \cdot \frac{a}{\sqrt{(1-a^2)}}$$

with

$$a = 1 - 2^{-(n+1)}$$

the normalized amplitude at $f_x = B$ for a permissible error of $-Q/2$. For example,

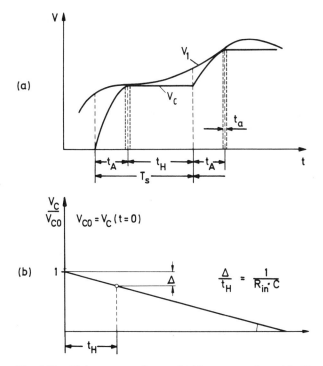

Fig. 5.21 Voltage waveforms inside a sample-and-hold circuit. (a) Sample-and-hold modes (simplified); (b) hold interval in more detail

if $n = 8$ bit, a bandwidth $f_3 = 16 \cdot B$ is needed. This formula is also valid for any amplifier with a first-order low-pass response in front of an A/D converter.[13]

If the time constant is no longer small compared with the acquisition time t_A; some averaging of the signal during sampling time t_A takes place. As a worst-case estimation, integration of the signal during t_A is assumed, i.e. $RC \gg t_A$, such that the signal can vary by a considerable number of quantizing steps without the capacitor voltage being able to track it. In this case, the frequency response can be approximated by the function

$$\frac{\sin \pi f_x t_A}{\pi f_x t_A}$$

This condition $t_A \ll RC$ usually holds for S/H within picosecond rise time sampling oscilloscopes. It is unusual for A/D converters. The frequency response of a S/H is treated in more detail in reference 14.

The sin x/x decrease due to the hold time t_H has an influence only if the S/H is used as a stand-alone unit. If a S/H is in front of an A/D converter, it is important that the S/H reaches and tracks the signal during acquisition time. An abrupt transition from sample to hold should occur when the sampling pulse changes state. This does not always happen in practice.

First, there is a certain delay between the sampling pulse and the actual sample-to-hold transition of the circuit. This delay can be divided into a constant so-called 'aperture delay time', which can be taken into account, and an aperture uncertainty time t_a which is changing from sample to sample because it depends at least on the signal level and the slew rate. This aperture uncertainty is difficult to measure and can only be estimated in many cases.[15,16]

The finite turn-off time of the switch inside the S/H, described by t_a, causes an error. In the worst case, this error will be the aperture uncertainty time t_a times the maximum slew rate of the signal. A more 'optimistic' estimation of the error uses the sin x/x function again with t_a now inserted. This seems to lead to an acceptable compromise between worst-case precision requirements and realistic circuit implementation.

The signal level attained at the end of the acquisition time t_A must be kept constant over the time interval t_H during which A/D conversion takes place. Assuming a sufficiently high OFF-resistance of the switch, then the only resistance in the discharge path of C is actually the amplifier input resistance R_{in}. If the capacitor voltage is approximated by the tangent to its decaying exponential function (Fig. 5.21(b)), then the permissible voltage droop Δ over the hold period t_H, assuming normalized values, is given by

$$\Delta \leq 2^{-(n+1)}, \text{ yielding } R_{in} \cdot C \leq \frac{t_H}{\Delta}$$

A further condition is obtained by considering the necessary decoupling between the signal source and the hold capacitor throughout t_H. Although a very high OFF-resistance of the switch is assumed, the actual isolation is not perfect due to the capacitor C_S shunting it (feedthrough). A capacitive voltage divider is therefore established. Thus by using the same $Q/2$ error criterion as before, we get

$$C_s/C \leq 2^{-(n+1)}$$

Other sources of errors are discussed in conjunction with the respective circuits in the following paragraphs. Especially in high-speed applications, the sampling time t_A is relatively long compared with the hold time t_H, and the circuit has to follow continuously the input signal before switched into the hold mode. Therefore the term 'track-and-hold circuit' is often used.

5.2.2 A Two-Phase Switch

In the auxiliary-charge-phase switch circuit, shown in Fig. 5.22, samples of the analog voltage V_1 are stored on the hold capacitor C_H. Due to diode D_1, the circuit can deal with positive voltages only. Therefore a bias voltage V_0 is added to allow for bipolar input-signals such that always

$$V'_s = V_1 + V_0 > 0$$

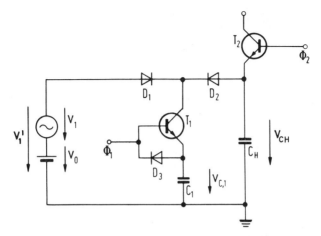

Fig. 5.22 Circuit of the two-phase switch

According to Fig. 5.23, the voltage ϕ_1 is negative and ϕ_2 is positive in phase I. The amplitude of ϕ_2 must be chosen such that $\phi_2 > V_1'_{max}$, where $V_1'_{max}$ is the maximum possible value of $V_1 + V_0$. Capacitor C_1 is charged via D_3 to the full value of ϕ_1 (neglecting the diode voltage drop). At the same time, C_H is charged to the voltage ϕ_2, if the base–emitter voltage of transistor T_2 is also neglected.

In the second phase, the voltages ϕ_1 and ϕ_2 are reduced to zero. Owing to the negative charge on C_1, diode D_3 is then reversed biased, whereas T_1 starts to conduct, thereby discharging both C_1 and C_H through D_2. Discharging continues until the condition $V_{CH} = V_1'$ is reached, where D_1 is turned on, allowing C_1 to be completely discharged. Due to their symmetry in the circuit, the voltage drops on

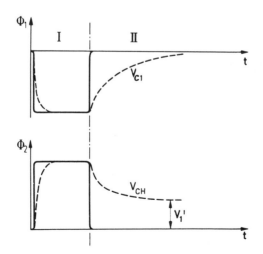

Fig. 5.23 Voltage waveforms corresponding to Fig. 5.22

D_1 and D_2 do not influence the result $V_{CH} = V_1'$ as long as they are equal. The hold-phase starts as soon as C_1 has been discharged, while D_1 and D_2, as well as T_2, are turned off.

Note that the discharging current for C_H is not supplied by the signal source but by the auxiliary circuit via T_1. In this way, relatively fast discharging of C_H can be achieved. This circuit was successfully used at the input of a bucket brigade (an analog shift register) at a clock frequency up to 30 MHz.[17] However, the circuit works only for positive signals.

5.2.3 The Diode Bridge Sample-and-Hold Circuit

The disadvantages of the preceding S/H circuit are overcome by the circuit shown in Fig. 5.24. This finds extensive use in high-frequency applications.

Let us first consider the hold mode that is controlled by the steady state of the two outputs of a symmetrical pulse generator. In this mode, the four diodes of the bridge, D_3, D_4, D_5, D_6, are reverse biased. The voltage of the hold capacitor C_H is sensed by a high-input impedance series feedback buffer amplifier. During the hold period, diodes D_1 and D_2 are conducting the constant current I_0 delivered by the two symmetrical current sources. In order to ensure the reverse biasing of the diodes D_3, ..., D_6, the cathode potential of D_1 must be lower than the lowest signal voltage V_1 while the anode potential of D_2 has to be higher than the maximum possible signal voltage V_1.

In the sampling mode, the pulse generator changes V_L and V_R such that D_1 and D_2 become reverse biased. If the capacitor voltage V_{CH} is larger than the output voltage V_1 of the pre-amplifier, then D_3 and D_6 are turned on to conduct the current I_0. Capacitor C_H is then discharged via D_6 and the current through D_3

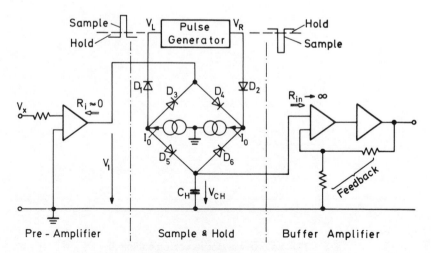

Fig. 5.24 Sample-and-hold circuit with a diode bridge

flows into the pre-amplifier output. Discharging continues until the two voltages V_{CH} and V_1 become equal. All the diodes of the bridge are then turned on to conduct the current $I_0/2$. Furthermore, the pre-amplifier is no longer loaded by the current I_0.

In the case that at the beginning of the sampling period the voltage V_{CH} is less than the signal voltage V_1, then diode D_5 conducts the current I_0 until balance is attained, while the pre-amplifier is the sink for the current I_0 of the right current source. All diodes are also conducting, as before, if $V_1 = V_{CH}$.

Charging and discharging of C_H is provided by the two current sources I_0. A linear rather than exponential voltage transient function, i.e. a constant voltage gradient dV_{CH}/dt, speeds up the response. Moreover, in the steady state of sampling the signal source is not loaded, so the voltage V_1 does not depend on the internal resistance. Compared with the two-phase circuit discussed above, there is no need for the initialization period I.

In addition, the charge on C_H, and hence its voltage, is changed by an amount corresponding to the difference between each two successive samples instead of being changed from a maximum value to the signal voltage. This property of voltage 'tracking' is suitable, especially when dealing with slower transient signals.

In an experimental realization of a diode-bridge S/H circuit, designed for video A/D converter operating at a sampling rate of 12 MHz, the following values were used:[16]

$$I_0 = 15 \text{ mA}, C_H = 100 \text{ pF, and } R_{in} = 1.5 \text{ M}\Omega$$

Thus, within an interval of $t_A = 20$ ns, a voltage change of 3 V is obtainable with the constant current charging/discharging, which is quite satisfactory for this video coding application. Charging of C_H by a constant current I_0 is achieved only if $|V_1 - V_{CH}| \gtrsim 400$ mV, i.e. as long as only two diodes are on. Considering the other case in which all the diodes of the bridge are conducting, the equivalent circuit of Fig. 5.18 still holds, i.e. voltages change exponentially. A charging RC-time constant of 2.7 nsec, determined by the diodes' I/V characteristic and the amplifier internal resistance R_i, is small enough compared with $t_A = 20$ ns. With an input resistance of the output buffer amplifier of $R_{in} = 1.5$ MΩ, the discharging time constant $R_{in} C_H$ is as high as 150 μs. This ensures that discharging during the hold period is negligible.

An important effect of S/H circuits is the time uncertainty (jitter) when switching from the sampling into the hold mode, leading to a corresponding uncertainty in the stored value V_{CH} depending, among other things, on the signal slew rate (aperture jitter). The trailing and leading edges of the pulses V_L and V_R, respectively, are illustrated in Fig. 5.25, in which a linear waveform for rise and fall is assumed.

As mentioned earlier, V_L and V_R bring the diodes D_1 and D_2 into the ON-state, respectively, thereby causing the currents I_0 to flow through the symmetric pulse generator. The sampling period is intended to terminate exactly at the beginning of the trailing edge of V_L and the leading edge of V_R, respectively. Assuming that $V_{CH} > 0$ and $V_1 \gtrsim V_{CH}$, then charging of C_H continues until V_L becomes less than

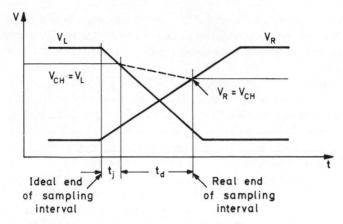

Fig. 5.25 Sample-and-hold turn-off waveforms (schematic)

V_{CH}, and hence I_0 flows through D_1. Since in this case $V_1 \approx V_{CH}$, D_4 and D_6 remain conducting until $V_R > V_{CH}$, when the discharging ceases.

Even with a fast rise and fall of V_L and V_R, a finite time interval known as the 'aperture time' elapses before the hold mode starts. As shown in Fig. 5.25, this period is composed of two intervals: the fixed aperture delay time t_d and the small fraction of time t_j that represents the time uncertainty or the variation of the delay time. This time t_j is accompanied by a corresponding amplitude uncertainty.

For the example considered, the measured aperture time is $t_d + t_j = 0.85$ ns, leading to an amplitude error of 5 quanta for a 9-bit resolution and a worst-case slope of $I_0/C_H = 0.15$ V/ns.

The isolation from the signal source in the hold mode is limited by the junction capacitances of the diodes D_3–D_6 and the ON-resistance of the conducting diodes D_1 and D_2 shunting the bridge centre to a.c. ground. This provides about 100 dB

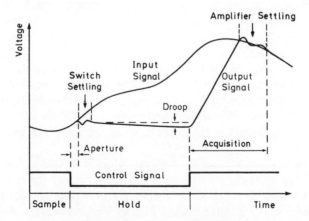

Fig. 5.26 Waveforms of a sample-and-hold circuit

attenuation, which is quite satisfactory. The isolation from the pulse generator is smaller (about 40 dB). However, it is less critical, since the resulting error occurs only during the remaining rise and fall times of V_L and V_R after the aperture time $t_d + t_j$. Furthermore, this error is reduced if the two diodes D_5 and D_6 are symmetrical.

A useful figure of merit for a S/H circuit is sometimes given by the ratio of the available charging current during the sample mode to the leakage current from the capacitor during the hold mode. This is approximately equal to the ratio of the slew rate to the droop rate at the output of the S/H.[18] Some of the parameters useful to describe S/H performance are summarized in Fig. 5.26.

5.2.4 Monolithic Sample-and-Hold Circuits

The diode bridge circuit sample-and-hold circuit of section 5.2.3 has become a standard hybrid circuit for high-speed applications. However, it is expensive. Therefore a number of attempts have been made to integrate MOS-transistors as switches together with the capacitor and surrounding amplifier circuitry.

A straightforward realization is shown in Fig. 5.27(a), where unity gain buffers are used to decouple the signal source from the capacitive load of the sampling

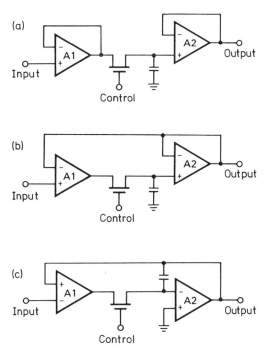

Fig. 5.27 Sample-and-hold amplifier configurations. (a) Buffered; (b) closed loop; (c) integrator

capacitor and to isolate the capacitor during the hold mode by the second buffer. However, the offset voltages of both buffers add up to an unfavourable value. The feedback system of Fig. 5.27(b) reduces the offset of the second amplifier by the gain of amplifier 1. The next improvement with respect to speed and accuracy is obtained by the circuit of Fig. 5.27(c), where the second amplifier A2 is connected as an integrator.[19]

The circuit implementation on a 6 mm² chip included silicon gate FETs as high-impedance amplifier input transistors and bipolar transistors which were mainly used in the rest of the amplifier circuitry. The slew/droop rate was

$$3\frac{V}{\mu s}\bigg/30\frac{mV}{s}$$

Acquisition/aperture delay times were 10 μs (20 V step)/80 ns, respectively, and the gain and non-linearity error was smaller than 0.01%. Circuits similar to the basic ones in Fig. 5.27 have been developed to improve performance or to have more functions for data acquisition on one chip.[20] The settling time to 0.01% of full scale is 4 μs. The specific problem of offset voltage error is solved by an auto-zeroing feature. p-channel FETs are used together with bipolar devices.

Another problem that is specific for MOS switches in sample-and-hold applications is capacitive feedthrough of the switch control signal to the hold capacitor. In reference 21 this has been solved by a dummy switch that is controlled by an opposite voltage and by adding a compensating capacitor. For offset compensation, an idle A/D conversion cycle is incorporated. Accuracy/resolutions are 8 bits/12 bits, respectively, settling time per one (out of eight) channels is 66 μs. A double polysilicon enhancement/depletion NMOS process with 6 μm line width features is used to make a 3.5 × 3.5 mm² chip.

5.2.5 Ultra-High-Speed Sample-and-Hold Circuitry

While hybrid sample-and-hold circuits are dominating the high-frequency domain with 13 bits resolution up to 10 MHz sampling rates, with 20 ps aperture uncertainty and a 25 ns acquisition time for 0.1% accuracy,[22] attempts are also being made to bring all the components onto one chip. A serious contender in this field is GaAs as a material and Schottky diodes or junction field effect transistors with a Schottky barrier, metal control gate, i.e. MESFETs as switches.[23] The basic switch consists of a bridge (Fig. 5.28) with four MESFETs instead of diodes. With a slew rate of 750 V/μs, an acquisition time of 2 ns, and an on-to-off ratio of 40 dB, sampling rates of up to 250 MHz are anticipated.

A recent approach has been developed for an ultra-fast transient digitizer (see section 3.6.8) for a 20 MHz sampling rate at 10 bits resolution.[24] The sample-and-hold circuit, on a ceramic substrate, consists of four integrated circuit chips, a preamplifier of 150 MHz full-power bandwidth, a monolithic diode bridge using MoAu Schottky diodes, a switch control circuit based on a trifilar transformer for symmetric control signals, and an output buffer amplifier. The complete S/H

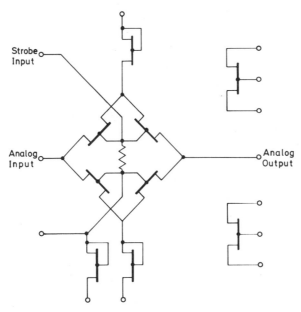

Fig. 5.28 Circuit of a GaAs sample-and-hold chip[23]

circuit typically has 80 MHz bandwidth in the tracking mode, minimizing gain and phase distortion within the Nyquist bandwidth of 10 MHz for the digitizer.

5.2.5 A Digital Sample-and-Hold Circuit

As pointed out in the preceding section, a high-quality S/H circuit must meet a number of requirements that are currently fulfilled only by expensive components. In view of the rapid developments in the field of digital integrated circuits, the trend is to move the interface between the analog and digital parts in an A/D converter, wherever possible, in favour of digital circuitry. One approach is the development of a digital S/H circuit, also known as 'sampling-on-the-fly'[25] or 'flash' type converter.

As shown in Fig. 5.29, the circuit is actually a parallel type A/D converter (section 3.2.2). The input voltage V_x is applied simultaneously to the $(m-1)$ comparators whose threshold voltages are provided by a uniform distance voltage divider in connection with V_{ref}. The input voltage V_x is usually a function of time, i.e. it is continuously changing. The comparators, which must be fast enough, are expected to follow the signal variations instantaneously. In this way, the momentary value of V_x is reflected in the outputs of the comparators, e.g. at that output where a transition from '0' into '1', or vice versa, takes place, and is shifted up or down in accordance with the analog input $V_x(t)$. The output of each comparator is then connected to a latch circuit with a sampling pulse applied to its control input. Upon the sampling command, the actual state of the comparators is stored in the latches. During the hold mode, the instantaneous amplitude is encoded by a suitable decoding logic circuitry for $\log_2 m$ bits.

Extension to higher resolution will be possible by using a second stage of the same type with an appropriate D/A converter in a post-subtraction structure, as shown in Fig. 5.29 with dashed lines. A delay element is necessary in the analog signal path in this case to compensate for the processing time of the first sub-circuit. The same delay must also be introduced to the sampling pulses applied to the next stage. The resulting aperture error of such a system is assumed to be considerable due to the signal amplitude dependent delays in the various parts of the circuit.

The main advantage of this system is that both functions of a S/H circuit and an A/D converter are integrated into one circuit. As in the case with parallel converters, the major drawback is the circuit complexity and hence cost. Moreover, due to the different propagation delays of the comparators, a considerable aperture error is to be expected.

An experimental circuit using a Read only Memory (ROM) as encoding logic is described in reference 26 for a 7-bit word length and a sampling rate of 30 MHz. The circuit was built using sixteen thick-film substrates (20 × 25 mm² each), each containing eight comparators and eight gates.

The high-speed potential of the digital sample-and-hold technique has been demonstrated by various designs. A 3-bit quantizer was incorporated in an experimental set-up for 200 MHz sampling rate.[26] Other designs are aimed at 4-bit

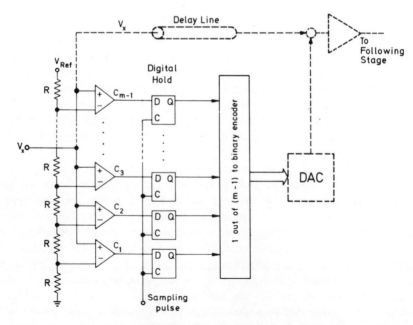

Fig. 5.29 A/D converter as a digital sample-and-hold

resolution and 100 MHz conversion rate for later extension to 8-bit.[27,28] All of them are based on very fast comparators with built-in latching functions that provide the digital sample-and-hold capability in a very fast speed range. This type of comparator is described in more detail in section 5.3.2.

Fully integrated versions for 9-bit resolution at 25 MHz word rate were commercially available by 1981.[29]

5.3 Comparators

5.3.1 Introduction

The function of a comparator is to deliver an output voltage which represents the result of a comparison between two voltages at its inputs. Comparators are therefore used in data converters to find the level of an input voltage that can be assigned to one out of m quantization levels. In fact, the number of comparators in a certain converter set-up is a direct measure of its complexity and cost. The circuit symbol of a voltage comparator is shown in Fig. 5.30(a).

The range of comparison of a voltage comparator $V_{min} \leq V_1 \leq V_{max}$ is shown on its input/output relationship in Fig. 5.30(b). Usually, $|V_{max}| = |V_{min}|$ or one of them is zero. The reference voltage V_{ref} (threshold voltage) can assume any value within this range. The output voltage V_2 takes only two levels (states). If

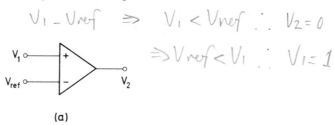

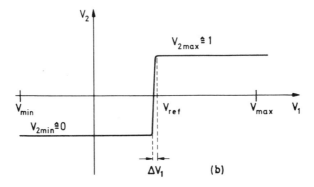

Fig. 5.30 A voltage comparator. (a) Circuit symbol; (b) transfer characteristic

$V_1 < V_{\text{ref}}$, then

$$V_2 = V_{2\min} \stackrel{\triangle}{=} \text{'0'}.$$

Also if

$$V_1 > V_{\text{ref}}$$

the output is

$$V_2 = V_{2\max} \stackrel{\triangle}{=} \text{'1'}.$$

The transition from logical '0' to logical '1' is not precise in practical comparators, but rather it has a finite width ΔV_1 of uncertainty due to the finite gain of the circuit (gain error). This width puts a limit on the number of admissible comparison levels (resolution) within the range $V_{\max} - V_{\min}$. Therefore it should be made as small as possible. Consider, for example, a 12-bit A/D converter. If the range $V_{\max} - V_{\min}$ is 10 V, then the corresponding quantum size is $Q = 10 \text{ V} : 2^{12} = 2.5$ mV.

The required amplification can be determined if the type of logic circuits is known (e.g. TTL, ECL, etc.) which the comparator has to drive. This determines the difference in level between '1' and '0' (for ECL it is about 0.8 V, for TTL about 3–5 V) from which the necessary amplification is calculated. For our example, the required amplification ranges between 320 for ECL and 1000–2000 in the case of TTL circuits.

Besides high amplification, a large bandwidth is also necessary in order to keep the transition time (response time) from one state to the other as small as possible. A measure of these two characteristics is the gain–bandwidth product $G \cdot B$. For those reasons, comparators have a structure similar to that of an operational amplifier. However, in contrast to wideband amplifiers, comparators are mostly operated in two extreme states, high- or low-level output, i.e. they are operated in their cut-off region and hence they are overdriven most of the time. Therefore the circuit should be designed to avoid adverse consequences of overdrive, mainly saturation of the transistors. Since comparators operate between these two extremes, the response time is more meaningful than the bandwidth. Also, in an analogy to amplifier terminology, the term 'input slew rate' (SR), which gives the maximum rate of change at which the input stage can follow without slew rate deterioration, is a useful specification. In addition, a high common-mode range is more important than for amplifiers, even if it is obtained at the expense of larger drift, smaller input resistance, and even instabilities within the linear region.

5.3.2 Bandwidth and Transient Characteristics

As mentioned earlier, the structure of a comparator is similar to that of an operational amplifier. On the other hand, the response time or delay is more important than the bandwidth, although they are related to each other, as we shall see below.

A simplified circuit diagram of a voltage comparator is shown in Fig. 5.31. This

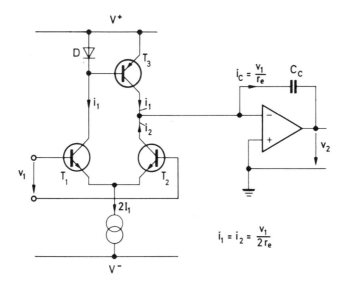

Fig. 5.31   Basic circuit diagram of a comparator

consists mainly of an input difference amplifier (T_1, T_2) and a current mirror (T_3, D) in the collector circuit which is then followed by an amplifier stage. In order to maximize the gain of the input stage, the load impedance of the transistors T_2 and T_3 should be as large as possible. In our case, the resistive load can be neglected, so that the maximum load is represented by the small capacitor C_C, often introduced for stabilizing the frequency response of the output amplifier.

From the circuit in Fig. 5.31 it follows that

$$\left|\frac{V_2}{V_1}\right| = \frac{1}{r_e \cdot \omega \cdot C_C}$$

with r_e the differential input resistance of a transistor in common-base connection. The gain–bandwidth product, defined by the frequency ω_1 at which $|V_2/V_1| = 1$ is thus given by $\omega = 1/(r_e C_C)$. Recalling that r_e is determined by the transistor input I/V characteristics and is given approximately by $r_e = V_T/I_e$ ($V_T \approx 26$ mV at room temperature) and that $I_e = I_1$ we get $\omega_1 = I_1/(V_T C_C)$.

If the input stage of the comparator is now overdriven, then only one of the transistors T_1 or T_2 is ON and carries an emitter current of $2I_1$. The current source guarantees the delivery of this current. The maximum charging current i_C of the capacitor C_C, is $2I_1$; hence the maximum possible slew rate of the output voltage is SR $= 2I_1/C_C$. Substituting ω_1 in this relation we get SR $= 2\omega_1 V_T$.

For example, consider an amplifier having a gain–bandwidth product $\omega_1/2\pi$ of 100 MHz, then the slew rate determined is about 33 V/µs. This means that for a voltage change of 5 V, as is the case with TTL-circuits, a rise time of about 150 ns is to be expected. In this example, the worst case, i.e. unity gain of the second amplifier, is considered. Otherwise, if higher amplification is provided, V_2 and the response time are correspondingly reduced.

Note that the relationship for the SR can also be used to determine the large-signal bandwidth, i.e. the frequency at which the amplifier is capable of amplifying a certain input voltage amplitude without distortion (without exceeding the maximum slew rate at the zero crossings of a sine wave).[30]

In high-speed comparators additional means are provided to improve the slew rate. Examples of these modifications are the use of cascade stages in conjunction with the difference amplifier at the input, amplitude clipping by Schottky-diodes, and/or the use of internal feedback to boost the input voltage.

A standard feature of high-speed comparators is the strobe, or latching capability. While a strobe facility forces the output of the comparator to become insensitive to the input signal and to assume one of its two logic output levels for further digital processing, the latching function locks the output in the logical state it was in at the instant the latch was enabled, allowing short input signals to be detected and held for further processing. Strobing is usually done close to the logic output of the comparator and latching is introduced in its input stage, and therefore signals only a few nanoseconds wide can be acquired and held with very low aperture time.

The principle of a latch circuit will be described relating to Fig. 5.32.[31] The differential amplifier input stage (standard in many fast comparator designs, some of them improving it by a buffer emitter follower at each input[32]) is extended by a pair of cross-coupled transistors T_3, T_4 connected to the collectors of T_1, T_2. The latch enable turns on the current of T_3, T_4 which is steered into the collector node of the differential pair that is lower in potential at that instant, thus introducing a

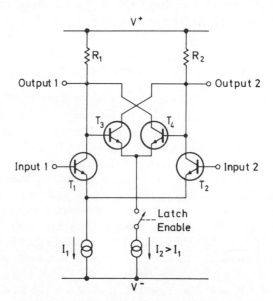

Fig. 5.32 Simple latch circuit inside a comparator

positive feedback action that latches the state of the outputs. In order to hold it, I_2 must be larger than I_1. This technique is employed in a number of comparator designs with further refinements to maintain or enhance speed in the tracking mode or to achieve low-power dissipation and/or input currents.[33,34]

The three most important speed parameters characterizing a sample-and-hold circuit are acquisition time, aperture time, and aperture uncertainty (jitter). It is interesting to compare these parameters for analog sample-and-hold circuits, as described in section 5.2.3, and for a latching comparator in a digital sample-and-hold scheme, as described in section 5.2.6.[35]

While large input voltage excursions increase the acquisition time of analog sample-and-hold circuits, they reduce it in the latching comparator due to overdrive of its input. Aperture delay and aperture jitter in analog S/H circuits are related to many parameters such as signal amplitude, slew rate, rise and fall times of control signals and circuit symmetry, whereas the latching function requires only a slight imbalance of the input signals to the differential amplifier. On the other hand, a comparator discriminates only whether a signal is smaller or larger than just one level, and therefore many of them are needed to quantize a signal within a specified range.

Therefore speed and resolution can be considered to be the most significant performance parameters of comparators which have to be traded against each other in a specific design. Table 5.2 gives an indication of these parameters for some well-known types of comparators. Accuracy is further determined by offset voltage and voltage drift, as well as by offset current, adding to the error budget. Common-mode voltage range should be large as well as bandwidth and stability in the tracking mode.

MOS-technology, while being suitable for high-density low-cost digital circuit functions, has a number of drawbacks for analog circuits. The absolute values of device parameters, such as threshold voltage which is important in differential amplifier configurations, varies by hundreds of millivolts, and tracking is within tens of millivolts. Therefore MOS-comparators are based on a.c. coupled amplifiers that are fed by a difference signal obtained by switching the signal and reference voltages to a flip-flop circuit that has been previously pulsed (precharged) to its unstable operating point. This type of latching circuit is extensively

Table 5.2 Performance parameters of available comparators

Type	Logic family	Input/output propagation Delay (ns)	Resolution (mV)
μA 710	TTL	40	1.5
Am 106	TTL	40	0.5
μA 760	TTL	25	0.5
NE 521	TTL	10	7.5
Am 685	ECL	6.5	2
SP 9750	ECL	3	5
MC 1651	ECL	2.5	25

used in dynamic memories as a sense amplifier.[36] Based on this technique, a 1 mV MOS-comparator has been built and has been shown to switch within 3 μs.[37]

In order to obtain good noise immunity, a MOS differential input amplifier has been analysed and tested. Latching was again obtained by a dynamic flip-flop. A sensitivity (resolution) of 0.5 mV has been achieved with offset voltage values of −0.48 mV mean value and 5.3 mV standard deviation. Switching takes place in 20 ns by means of a strobe amplitude of 5 V. The chip area is 0.05 mm² when producing the circuit in a standard silicon-gate NMOS-technology.[38]

5.3.3 Operational Amplifiers as Comparators

As stated earlier, operational amplifiers, which are designed as wideband amplifiers, can also be used as comparators.[39] In this section some technical problems, as well as ways to overcome them, are discussed. Figure 5.33(a) illustrates an operational amplifier with a Zener diode Z_F in its feedback loop and a resistor R at its input. The Z-diode, whose I–V characteristic is shown in Fig. 5.33(b), protects the amplifier against overloading. In this case, V_Z (reverse

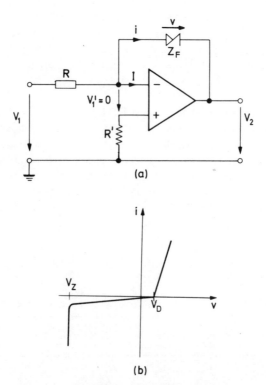

Fig. 5.33 Operational amplifier as a comparator. (a) Circuit diagram; (b) characteristic of the Z-diode

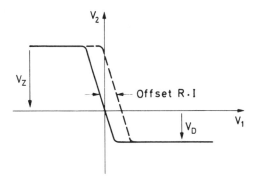

Fig. 5.34 Input/output characteristic of the comparator circuit in Fig. 5.33

breakdown voltage) must be smaller than the overload limit of the amplifier's output voltage. For large negative values of V_1, breakdown occurs at V_Z due to the amplifier's phase inversion and to the fact that the input voltage V_1' of the amplifier is $V_1' \approx 0$ (virtual ground); therefore $V_2 = V_Z$. For large positive values of V_1, the output voltage V_2 is $V_2 = V_D$ (V_D is the forward voltage of the diode). With zero input current (ideal condition), the transfer characteristic (Fig. 5.34) has no offset (i.e. $V_2 = 0$ for $V_1 = 0$) and a slope of R_F/R, where R_F is the static differential impedance of the diode at $V = 0, I = 0$.

In fact, a small offset current I will flow into the input of an operational amplifier due to the base currents of the input transistors. This offset current results in a non-zero input voltage ($V_1 = I \cdot R$) even if V_2 and hence I are zero. The transfer function is correspondingly shifted to the right by the value $I \cdot R$ (shown dashed in Fig. 5.34). A compensation of this offset is possible if a resistor R' is connected in series with the non-inverting input of the amplifier. In the case of equal offset currents of both inputs, then $R' = R$. With $V_1 = 0$, both inputs are biased accordingly by an amount of $R \cdot I$ V, leading to $V_1' = V_2 = 0$. Moreover, additional voltage offsets can be compensated by varying R' carefully. The main drawbacks of this circuit are its finite input impedance and limited noise immunity, characterized by undesired switching caused by noise near the transition region.

If a comparison with a voltage different from zero is required, a circuit as shown in Fig. 5.35 can be used. Overload protection is provided by the two shunt

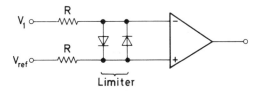

Fig. 5.35 Comparator circuit for a non-zero reference

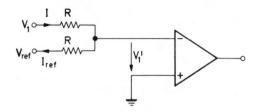

Fig. 5.36 Comparator with improved Common Mode Rejection

diodes. Here the reference voltage V_{ref} is applied to the non-inverting input. The output voltage transition at $V_1 = V_{ref}$ exhibits a slope that is limited by the open-circuit amplification of the operational amplifier.

Due to the finite common-mode rejection, the transition is shifted. This drawback can be eliminated by the modification shown in Fig. 5.36 such that the transition for the amplifier takes place at $V_1' = 0$.

Finally, the Schmitt trigger circuit (regenerative comparator), shown in Fig. 5.37, is presented. This circuit has the advantage that unwanted transitions, e.g. due to noise voltages, are eliminated by an adjustable hysteresis in its transfer function. Also, due to an inherent positive feedback, large transition slopes are attainable.

Positive feedback is provided by R_2 and R_3 together with the usual negative feedback. The two back-to-back connected Z-diodes provide voltage-limiting in both directions due to the symmetrical I–V curve shown in Fig. 5.38.

Assuming ideal operational amplifier characteristics, i.e. infinite gain ($V_1' = 0$ for a finite value of V_2) and negligibly small input currents to both input terminals, then

$$V_{R2} = V_F$$
$$I_{R2} = I_{R3} = V_F/R_2$$
$$V_{R3} = I_{R3} \cdot R_3 = V_F \cdot R_3/R_2$$
$$V_2 = V_{R3} + V_F = V_F(1 + R_3/R_2)$$

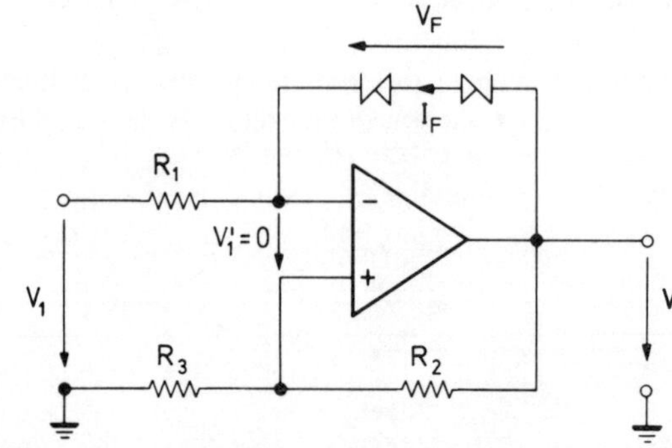

Fig. 5.37 Schmitt trigger

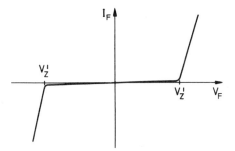

Fig. 5.38 Characteristic of the feedback path

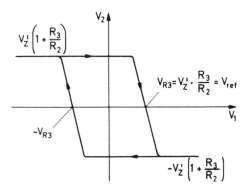

Fig. 5.39 Transfer characteristic of the Schmitt trigger

For large negative input voltages V_1, then $V_F = V_Z'$, so that $V_2 > 0$. Similarly, for large positive values $V_F = -V_Z'$, leading to a negative output voltage V_2 which is independent of V_1. The resulting input/output characteristic, shown in Fig. 5.39, exhibits an hysteresis. Starting with negative values of V_1, then V_{R3}, given by the equation above, presents a positive reference. On the other hand, if V_1 is made positive, V_{R3}, and therefore the reference, changes sign. Thus for decreasing V_1, the transition voltage is negative, and vice versa. The width of the hysteresis can be varied by R_2, R_3 according to the degree of noise immunity required. The transition region is very narrow due to positive feedback. An ultra-high-speed version of a Schmitt trigger circuit with 200 ps rise time and controllable hysteresis can be achieved.[40]

5.3.4 Tunnel Diode Comparators

Tunnel diode comparators are the fastest among all comparators operating at room temperature. They have been used in several high-speed A/D converter

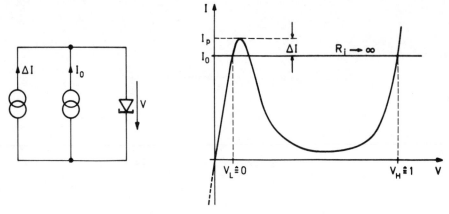

Fig. 5.40 I–V characteristic of a tunnel diode operating as comparator

designs since they provide the minimum switching time down to 30 ps (1 ps = 10^{-12} s).

A tunnel diode is, in fact, a current comparator whose reference is given by its peak current I_p (Fig. 5.40). This value varies from type to type (typically, 1 mA $\leq I_p \leq$ 100 mA). The actual reference current can be changed by adding a bias current I_0. The effective reference current is modified to become $I_p - I_0$. Maximum switching speed is obtained if the comparator is driven from an ideal current source, i.e. if $R_i \to \infty$. In this case, the load line, as shown in Fig. 5.40, is a horizontal line.

For an input current $I > I_p - I_0$ the diode is switched from its low-voltage state V_L to its high-voltage state V_H. (V_L = 50 mV, V_H = 500 mV for germanium. These values are approximately doubled in the case of GaAs diodes.) The switching time is given approximately by

$$t_r = \frac{\Delta V}{0,8 \cdot I_p}; \Delta V = V_H - V_L$$

with I_p/V_p the peak current/voltage. This means that the rise time is equal to the time required to change the voltage on the diode capacitance C by $\Delta V = 0.5$ V through the current $0.8 \cdot I_p$, assuming a minimum overdrive of 10%.

5.3.5 The Josephson Junction as a Comparator

Josephson junctions are ultra-fast, low-power devices exhibiting threshold characteristics (Fig. 5.41) at cryogenic temperatures.[41] They consist of superconduction connection lines (Pb alloys) separated by a thin oxide layer (junction). This layer is superconducting when starting from $I = 0$ up to a certain threshold value I_{max}, where the conduction mechanism changes to assume a finite voltage of about 2.5 mV. I_{max} depends, in addition, on a magnetic field that can be

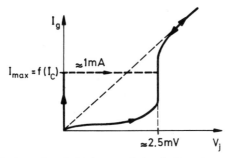

Fig. 5.41 *I–V* characteristic of a Josephson junction

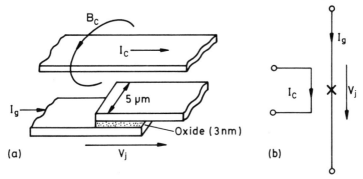

Fig. 5.42 Josephson junction. (a) Structure, (b) symbol

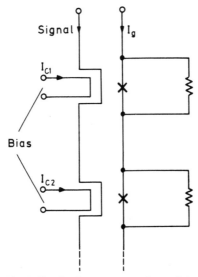

Fig. 5.43 Basic structure of parallel A/D converter based on Josephson junctions

applied to the junction via a control line running on top of the junction (Fig. 5.42(a)). A second control line can provide a bias field or act to establish a two-input gate function.[42] By means of the control lines the junction can operate as a controlled switch, i.e. as a comparator. The switching takes place very quickly, and switching times in the order of 30 ps have been measured. The device itself should be capable of switching to within approximately 1 ps.

Basically, a very simple parallel converter can be designed by extending the number of junctions through series connection, with each device subjected to the signal but at a different bias level,[43] as shown in Fig. 5.43.

5.4 Voltage Reference Circuits

Voltage references are essential for the operation of data converters. For example, they represent, in conjunction with comparators, the core of any A/D converter. Precision and long-term stability are the foremost design criteria, since voltage references determine the accuracy and stability of converters. Therefore they must deliver a substantially constant voltage over both time and temperature.

5.4.1 A Zener Diode Reference Circuit

One of the commonly used reference circuits is the compensated Zener reference diode with a buffer circuit (Fig. 5.44). Usually, the so-called compensated sub-surface (or buried) Zener diodes are used. These devices undergo an avalanche breakdown that occurs under the surface of the semiconductor, thereby giving better long-term stability and noise performance compared with the traditional surface breakdown diodes. Temperature compensation is achieved by connecting two diodes back-to-back, i.e. a forward biased diode in series with the reverse biased Zener diode. Thus if both diodes have approximately equal and opposite

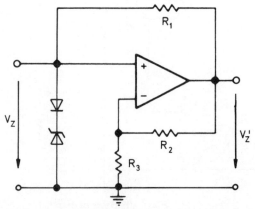

Fig. 5.44 A temperature-compensated Zener reference circuit

temperature coefficients, a temperature-stable voltage is obtained (tempcos ranging from 100 ppm/K to less than 1 ppm/K are obtainable). Further, magnification of this voltage (usually about 6.4 V) is possible if the buffer circuit, shown in Fig. 5.44, is designed accordingly. To yield a stable voltage V_Z the diodes are connected via resistor R_1 to the output of the operational amplifier instead of to the power supply.[44] The resulting reference voltage is then $V_Z' = V_Z(1 + R_2/R_3)$.

5.4.2 The Bandgap Reference

This reference circuit, shown in Fig. 5.45, was initially proposed by Widlar.[45] The name relates to the value of the reference voltage which is close to the value V_{G0}, that is, the silicon pn-junction voltage extrapolated to the absolute temperature $T \to 0$, which is $V_{G0} = E_{G0}/e = 1.205$ V, where e is the electron charge and E_{G0} the energy bandgap of silicon.

The circuit is based on the compensation effect of two temperature dependent voltages of opposite temperature gradient. In the circuit of Fig. 5.45, the base voltage V_{BE} of T_1 has a negative temperature coefficient, whereas the current through R_1 and hence the voltage V_{TK} across it increases with temperature. By suitable choice of the weighting factors, a first-order compensation is achieved.

An ideal operational amplifier keeps its input difference voltage at zero; therefore $I_{C1} = I_{C2}$ and hence $I_{E1} = I_{E2}$. It is essential to note that the emitter areas of T_1 ($\hat{=} A_1$) and T_2 ($\hat{=} A_2$) must be different (if T_1 and T_2 are equal, their

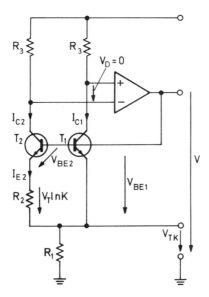

Fig. 5.45 The bandgap reference after Widlar[45]

collector resistors and their currents must be different), such that

$$V_{BE1} = V_{BE2} + R_2 I_{E2}$$

and

$$R_2 I_{E2} = V_{BE1} - V_{BE2} = V_T \cdot \ln \frac{A_2}{A_1} = \frac{kT}{e} \cdot \ln \frac{A_2}{A_1}$$

where

T = absolute temperature

k = Boltzmann's constant

$$\frac{kT}{e} = 25.9 \text{ mV at room temperature}$$

$$V_{TK} = 2R_1 \cdot I_{E2} = 2 \frac{R_1}{R_2} \cdot V_T \cdot \ln \frac{A_2}{A_1} \sim T$$

From the circuit, the output voltage is

$$V = V_{BE1} + V_{TK}$$

The relation for V_{BE1} has to be taken from semiconductor *pn*-junction theory[46] and its derivative added to the derivative of V_{TK} to get a zero total temperature dependence yielding

$$V(T_0) = V_{G0} + \left(3 - m + \frac{T_0}{R_2} \cdot \frac{dR_2}{dT}\bigg/_{T=T_0}\right) \cdot V_T(T_0)$$

where

V_{G0} = extrapolated bandgap voltage of silicon

m = correction factor: $1 \leq m \leq 1.5$,

giving approximately

$$V(T_0) \approx V_{G0} + 2V_T = 1.205 \text{ V} + 52 \text{ mV} = 1.257 \text{ V}$$

a value that is close to V_{G0}. The operational amplifier in the circuit can be used to amplify the voltage V by connecting T_1 via a voltage divider to the output. Devices are available that exhibit temperature stability of 10 ppm/K and 25 µV/month long-term stability. They can be operated from low-voltage power supplies over a wide temperature range.

5.5 Sawtooth Generators

5.5.1 Introduction

A sawtooth wave period consists of two unsymmetrical segments. The voltage in the first period increases linearly with time (trace period) and then drops suddenly

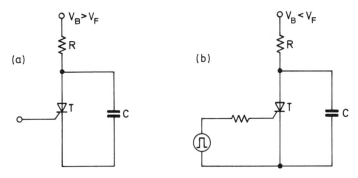

Fig. 5.46 A sawtooth generator using a thyristor. (a) Free running; (b) external trigger

in (ideally) zero time during the second period (flyback period). As mentioned in section 3.5, this waveform is used in indirect converters to convert the analog amplitude into a proportional time interval. Sawtooth generators are also essential components in oscilloscopes, television sets, and recording equipment.

Sawtooth generators are classified mainly into triggered and free-running types, depending on whether external trigger pulses are necessary for their function. Furthermore, a combination of the two operating modes (i.e. externally triggered rise with an internally controlled flyback) are common.

A free-running sawtooth generator, in its simplest form, is shown in Fig. 5.46(a), using a thyristor switch. Upon applying the supply voltage V_B, which is larger than the firing voltage V_F of thyristor T, capacitor C begins to charge through R. Charging continues until the capacitor voltage V_C reaches V_F, where the thyristor turns on. Through the on-resistance of the thyristor, capacitor C is then discharged until $V_C < V_H$ (the holding voltage of T). In this case, the switch is opened again and a new cycle is started. The repetition frequency is determined mainly by the RC time constant. The resulting waveform is composed of two sections of exponential rather than linear functions.

If, on the other hand, V_B is smaller than V_F, as in Fig. 5.46(b), then capacitor C charges exponentially towards V_B until a control voltage (trigger pulse) is applied to the thyristor's gate. This causes the thyristor to fire, thereby discharging C. The linearity of the obtained waveform may be improved if the resistor R is replaced by a transistor to supply a constant, rather than exponential, charging current.

5.5.2 The Miller Integrator

If a voltage step of amplitude V_S is applied at time $t = 0$ to an uncharged capacitor C which is connected in series with a resistor R, then the capacitor voltage is given by

$$V_C = V_S \left(1 - e^{-t/RC}\right)$$

This voltage function corresponds to an exponential current in the circuit of the

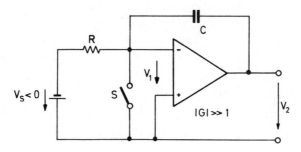

Fig. 5.47 Basic circuit of a Miller integrator

form

$$i = \frac{V_S}{R} e^{-t/RC} = \frac{V_S}{R} - \frac{V_C}{R} = I_0 - \Delta i$$

where

$$I_0 = i(t=0)$$

This means that the non-linear capacitor voltage change is due to a non-constant charging current. The decrement Δi stands for the deviation of i from its starting value I_0, and therefore it is used as a measure for the non-linearity error expressed as $\Delta i/I_0 = V_C/V_S$. From this formula it follows that this error is small for small values of V_C, i.e. if only a small part of the exponential curve is used. Miller integrators are based on this conclusion.

Figure 5.47 illustrates the basic circuit of a Miller integrator using an operational amplifier. Before starting, the switch S is closed so that $V_1 = 0$ and a current $I_0 = V_S/R$ flows. If S is then opened, I_0 charges C (assuming infinite input impedance of the amplifier). Assuming that the gain of the amplifier is high, i.e. $G = |V_2/V_1| \to \infty$, then $V_1 \to 0$ for finite values of V_2. Hence, the charging current remains constant. In fact, G is not infinite, leading to non-zero V_1. This, in turn, causes the charging current to vary by a quantity $\Delta i = V_1/R$. Relating this error to the initial current, the associated linearity error is given by

$$\frac{\Delta i}{I_0} = \frac{V_1}{R} \cdot \frac{R}{V_S} = \frac{V_2}{G \cdot V_S}$$

It becomes obvious that for the same values of $V_C = V_2$ and V_S as before, the linearity error is reduced by a factor of G (the gain of the operational amplifier).

5.5.3 The Bootstrap Integrator

The main disadvantage of the Miller integrator, namely its still exponential trace voltage, may be completely avoided if the charging current is kept constant. A circuit that meets this condition is shown in Fig. 5.48. At $t = 0$, switch S is opened so that capacitor C starts charging by a current $I = V_S/R$. If the increasing capacitor voltage is then added to the battery voltage V_S via a buffer stage having

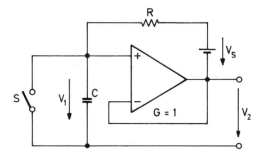

Fig. 5.48 A simplified bootstrap circuit

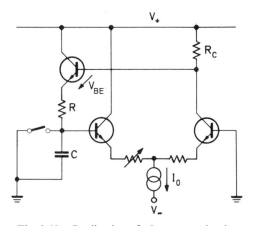

Fig. 5.49 Realization of a bootstrap circuit

a gain $G = 1$, the voltage drop across R and consequently the charging current remains unchanged. This leads to

$$dV_C/dt = V_S/RC = \text{constant}$$

i.e. the voltage V_C is a linear function of time. As shown in Fig. 5.48, the constant voltage source V_S is floating (i.e. none of its terminals is grounded); hence it is not easy to be implemented practically. One way of doing this is to use a differential amplifier. As illustrated in Fig. 5.49, we get

$$V_S = V_+ - R_C \cdot I_0/2 - V_{BE}$$

Assuming a constant V_{BE} of 0.6 V and a constant current I_0, supplied by the constant current source, then the voltage V_S remains substantially constant. The maximum capacitor voltage and hence the maximum sawtooth amplitude can be varied via the current I_0.

5.5.4 Switched Constant Current Charging

The circuit described here delivers, through a constant charging current of a capacitor, a linear sawtooth waveform whose amplitude as well as trace and

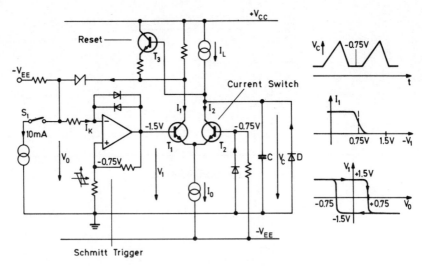

Fig. 5.50 A triggered sawtooth generator

flyback periods are variable. As shown in Fig. 5.50, it consists of a Schmitt trigger (section 5.3.3) and a current switch. While triggering of the circuit is controlled externally by the switch S_1, resetting is internally started via T_3 as soon as the capacitor voltage V_C reaches a certain (adjustable) value.

In the quiescent state with positive input voltage $V_0 > 0$, the output of the Schmitt trigger is negative, causing T_1 to be cut off. At the same time, T_2 is conducting, so that $I_2 = I_0$. Furthermore, with $I_2 > I_L$, diode D is conducting, thereby making $V_C \approx -0.75$ V. If S_1 is now closed for a short time, the Schmitt trigger changes state to give a positive output voltage V_1. T_1 is accordingly turned ON and T_2 stops conducting, yielding $I_1 = I_0$. This state is held dynamically, even if S_1 is opened. The current I_L, now supplied by the current source, charges 'linearly' the capacitor C. Transistor T_3 (reset transistor) starts to conduct when V_C becomes larger than the collector voltage of T_1 and resets the Schmitt trigger to its original state. Thus, T_2 becomes ON again (i.e. $I_2 > I_L$), thereby forcing C to discharge. At $V_C = -0.75$ V discharging ceases.

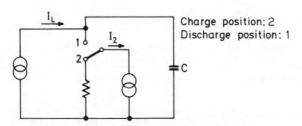

Fig. 5.51 Simplified equivalent circuit of Fig. 5.50

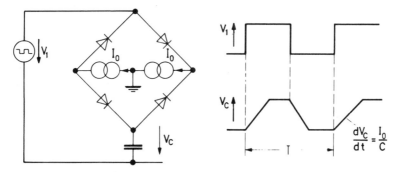

Fig. 5.52 Generating linear ramps

Although V_C is a function of time, the slope of V_C, as well as its amplitude, are determined by the static parameters of the circuit, e.g. resistors and voltages. Only S_1 (triggering) needs to be closed long enough to allow the Schmitt trigger to change state. A simple equivalent circuit for charging and discharging of the capacitor C is given in Fig. 5.51.

The circuit in Fig. 5.50 represents one example for the implementation of the more fundamental circuit of Fig. 5.51. A standard way of implementing constant current sources is by transistors with series feedback or switched current sources by means of a differential amplifier configuration (current switch, section 5.1.5), where a control voltage of ± 200 mV is satisfactory.

An alternative method for triggering and resetting the charging of a capacitor is to use the balanced diode bridge sample-and-hold circuit described in section 5.2.3. As illustrated in Fig. 5.52, a square-wave generator is applied to the diode bridge causing I_0 to charge, at a constant rate, capacitor C towards the voltage V_0. Through appropriate selection of I_0, the time interval T, and the amplitude V_0 of the square wave it is possible to obtain triangular, trapezoidal, and sawtooth

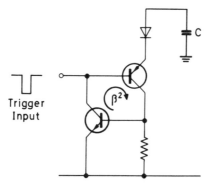

Fig. 5.53 Circuit used for minimizing fly-back period

waveforms. In the last case, however, it is usually desired to have a small discharge period compared with the charging one; hence the use of a constant current source seems to be unsuitable. Therefore a circuit as shown in Fig. 5.53 can be used to accelerate the discharging. Due to the existing positive loop gain within the two transistors (as in a thyristor), fast discharging is guaranteed. This is initiated either with an external trigger or when voltage V_C exceeds a certain threshold voltage.

5.5.5 Staircase Generators

Since the resolution of an A/D converter is determined by its quantum size Q, a staircase voltage having a step-size Q can be a replacement for the sawtooth voltage in indirect converters.

In a feedback A/D converter (section 3.42), the counter triggers every clock pulse a one-shot multivibrator to deliver an impulse of duration t_0. This impulse drives a constant current I into a capacitor such that its voltage is successively increased by an increment $V = I \cdot t_0/C$, providing the required reference. A buffer stage is usually connected before applying the capacitor voltage to the comparator. Resetting of this function can be made either internally or externally. A good linearity over the whole dynamic range is obtainable.

Another simple staircase generator, also called a diode pump, is shown in Fig. 5.54 using two FET-input operational amplifiers. A positive pulse V_{in} charges capacitor C via the resistor R and diode D_1 by a certain amount. If the input voltage V_{in} switches back to zero, the voltage on the capacitor C drives the cathode of diode D_2 negative and C discharges, while C_F becomes charged by a

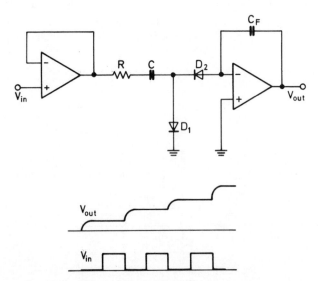

Fig. 5.54   Staircase generator based on charge pumping

specified amount to produce one step of the staircase output waveform. After discharge of C, D_2 stops conducting and C_F holds the voltage.

5.6 Counters

Counters are fundamental components in several data converters such as servo, indirect, and non-linear A/D converters. They are available as standard building blocks in integrated circuit form. In fact, one of the earlier conversion techniques is based on only counting step by step in a feedback structure (section 3.4.2).

5.6.1 Asynchronous (Ripple) Counters

The basic building block of a ripple counter is the multi-purpose JK flip-flop, whose symbol and truth tables are given in Fig. 5.55, where J and K are the date inputs and C is the clock input. The complementary output is obtained at the terminals F and $\bar{F}$, respectively. By means of the two inputs $\bar{R}$ and $\bar{S}$ it is possible to initiate the output to assume any required state. For example, with $\bar{S} = $ 'L' the output F = 'H', while if $\bar{R} = $ 'L', F = 'L'. This means that $\bar{R}$ resets the counter to zero, i.e. it is the erase or reset input. Moreover, the flip-flops are assumed to be of the master–slave type, in which the output changes state as soon as the clock changes from H to L ('1' to '0'). The storage property of this stage shows itself in its truth table. The state $F(k + 1)$ at the instant $k + 1$ is shown to depend on the preceding state $F(k)$.

The truth table indicates that the flip-flop (FF) changes state and 'toggles' each clock pulse if both inputs J and K are 'High', i.e. it functions in this case as a T-type flip-flop (Trigger-FF). This property is used in asynchronous binary counters, as shown in Fig. 5.56 for the case of a 4-bit counter. Four FF-stages are connected in cascade with the F-output of each stage which is connected to the clock input of the following stage. The pulses to be counted are applied to the clock input of FF_0 the first flip-flop. All the J and K inputs are connected to the supply voltage.

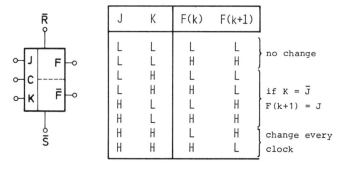

J	K	F(k)	F(k+1)	
L	L	L	L	no change
L	L	H	H	
L	H	L	L	if K = $\bar{J}$
L	H	H	L	F(k+1) = J
H	L	L	H	
H	L	H	H	
H	H	L	H	change every clock
H	H	H	L	

Fig. 5.55 Circuit symbol and truth table of a JK flip-flop

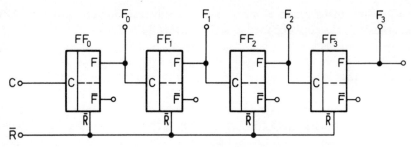

Fig. 5.56 A 4-bit ripple counter

Before counting is started, all stages are cleared (reset) by making $\bar{R} := $ Li during counting it is $\bar{R} := H$. Figure 5.57 indicates the waveforms at various locations in the counter. The number associated with each bit (weight) indicates its significance in the 4-bit straight binary code word. The input clock C assumes 16 pulses before the pulse pattern repeats. From the table it is obvious that the frequency of the output F_0 is twice that of F_1 and so on. The chain therefore acts also as a frequency divider with a division ratio of $2^4 = 16$. Each row in the table also gives the binary coded value of the number of pulses received up to this instant.

The drawback of this type of counter is the fact that the settling time (carry propagation delay) depends upon the number of stages undergoing a change of state between consecutive clock pulses. Consider, for example, the extreme condition at the instant of transition from 15 to 16, whereby all the four stages change state. In this case, FF_0 starts to assume the state '0', causing the change of state of

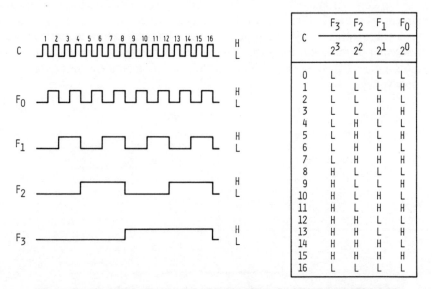

Fig. 5.57 Timing diagram and truth table of the ripple counter in Fig. 5.56

FF_1 and so on. Since these transitions occur sequentially, the total propagation delay is the sum of the individual propagation delays of the stages. The maximum input frequency is therefore limited by this value. It decreases with an increasing number of stages, making this counter structure unsuitable for many applications. Therefore if high-speed counting is required, synchronous counters have to be used.

5.6.2 Synchronous Counters

In a synchronous counter the maximum input frequency depends on the propagation delay of a single flip-flop only. This is because all stages are clocked simultaneously in parallel from the same source (Fig. 5.58).

The state of the inputs J and K determine whether this stage changes state upon receiving the next clock or not. This kind of logic capability is employed for anticipation of switching. From the table of Fig. 5.57 it follows that FF_1 changes state only if FF_0 assumes the state H. This implies that the inputs J_1 and K_1 are connected in parallel to F_0. Correspondingly, FF_2 changes state only if both F_0 and F_1 are high. In this stage the two input pairs J_1, K_1 and J_2, K_2 are therefore connected, as shown in Fig. 5.58. Also, in the next stage (FF_3) one more condition has to be met for switching, etc.

If instead a modulo-m counter is required rather than the modulo-2 counter just described, then at the clock pulse $m-1$, i.e. the state immediately before the required resetting, a logic circuit is used to decode that condition and to make all the J-inputs to 'L' and all K-inputs to 'H'. In this way, all stages are cleared at the next clock pulse as required and a new counting cycle is started.

5.6.3 Up/Down Counters

For the charge balancing converter (section 3.5.3), for instance, an up/down counter is required. Considering again the table in Fig. 5.57, the forward counting

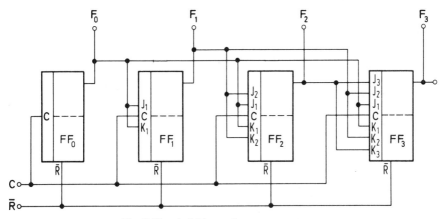

Fig. 5.58 A 4-bit synchronous counter

is characterized by the fact that each flip-flop undergoes a change of state and only if the preceding stages change from 'H' into 'L'. Alternatively, the counter can reverse its direction of counting if a change of state of a certain stage from H to L takes place and if all preceding stages are in the 'L' state. This simply means that backward counting could be affected if each output $\bar{F}_k$ instead of F_k is used to trigger the succeeding flip-flop FF_{k+1}.[44] By using suitable switches for choosing the logic required, the direction of counting can be controlled externally.

References

1. Ebers, J. J., and Moll, J. L., 'Large-signal Behavior of Junction Transistors', *Proceedings IRE* **42**, 1761–72, December (1954).
2. Gray, P. E., et al., *Physical Electronics and Circuit Models of Transistors*, SEEC, Vol. 2, Wiley, New York (1964).
3. Köhler, H., and Weidner, H., 'Ein schneller Spannungsschalter für Digital-Analog-Wandlung', *Elektronik* **17**, 17–19, January (1968).
4. '12 Bit Hybrid D–A Converter uses Monolithic Design', *Electronics* **54**, 249, September (1981).
5. Saul, P. H., 'Successive Approximation Analog-to-digital Conversion at Video Rates', *IEEE Journal* **SC-16**, 147–151, June (1981).
6. Cobbold, R. S. C., *Theory and Applications of Field-effect Transistors*, Wiley Interscience, New York (1970).
7. Reiniger, K. D., and Tränkler, H. R., 'Digital-Analog-Umsetzer mit FET-Schaltern', *Electronik* **21**, 39–43, February (1972).
8. *Data Acquisition Handbook*, Intersil Inc. (1980), Chapter 6.
9. Bernstein, H., 'Analogschalter, Theorie und Anwendungen', *Elektronikindustrie* **1974**, 33–6, March (1974).
10. Kitsopoulos, S. C., and Strauss, R. W., 'Transmission Gates with Negative Feedback', *1970 ISSCC Digest of Technical Papers*, 154–5.
11. Zuch, E. L., 'Pick Sample-holds by Accuracy and Speed and Keep Hold Capacitors in Mind', *Electronic Design* **26**, 84–90, December 20 (1978).
12. Taub, H., and Schilling, D. L., *Principles of Communication Systems*, McGraw-Hill, Tokyo (1971), Chapter 5.
13. Connolly, J. J., Rosenbaum, M. H., and Rittenhouse, L., 'Critical Design Parameters and Test Methods for Ultra High Speed Analog-to-digital Conversion', *Journées d'electronique* **73**, Lausanne, 16–18 October 1973, *Digest of Techcical Papers* 373–411.
14. Barbieri, G. G., Cominetti, M., and D'Amato, P., *The Sample and Hold Process in a PCM Video System*, British Post Office, Research Department Library, Translation No. 2961.
15. Johnston, R., 'Analyzing the Dynamic Accuracy of Simultaneous Sample-and-Hold Circuits is Straightforward', *Electronic Design* **19**, 80–83, September 13 (1973).
16. Gray, J. R., and Kitsopoulos, S. C., 'A Precision Sample and Hold Circuit with Sub-nanosecond Switching', *IEEE Transactions* **CT-14**, 389–96, September (1964).
17. Sangster, F. L. J., and Teer, K., 'Bucket-brigade Electronics—New Possibilities for Delay, Time-axis Conversion, and Scanning', *IEEE Journal* **SC-4**, 131–6, June (1969).
18. Zuch, E. L., 'Keep Track of a Sample-hold from Mode to Mode to Locate Error Sources', *Electronic Design* **26**, 80–87, December 6 (1978).

19. Stafford, K. R., et al., 'A Complete Monolithic Sample/hold Amplifier', *IEEE Journal* **SC-9**, 381–7, December (1974).
20. Gasparik, F., 'An Autozeroing Sample and Hold IC', *1980 ISSCC Digest of Technical Papers* 132–3.
21. Bienstman, L. A., and De Man, H., 'An 8-channel µP Compatible NMOS Converter with Programmable Ranges', *1980 ISSCC Digest of Technical Papers* 16–17.
22. *Data-Acquisition Databook 1982*, Vol. II, 14/15–14/18, Analog Devices Inc, Norwood, Mass. (1982).
23. Saul, P. H., 'A GaAs MESFET Sample and Hold Switch', *IEEE Journal* **SC-15**, 282–5, June (1980).
24. Muto, A. S., et al., 'Designing a Ten-Bit, Twenty-Megasample-per-Second Analog-to-Digital Converter System', *Hewlett-Packard Journal* **33**, No. 11, 9–20, November (1982).
25. Bernina, D., and Barger, J. R., 'High-speed, High-resolution A/D Converter: Here's How', *EDN* **18**, 62–6, June 5 (1973).
26. Woodward, Ch. E., Koukle, H. H., and Naiman, M. L., 'A Monolithic Voltage-comparator Array for A/D Converters', *IEEE Journal* **SC-10**, 392–9, December (1975).
27. Nordstrom, R. A., 'High Speed Integrated A/D Converter', *1976 ISSCC Digest of Technical Papers* 150–51.
28. Saul, P. H., Fairgrieve, A., and Fryers, A. J., 'Monolithic Components for 100 MHz Data Conversion', *IEEE Journal* **SC-15**, 286–90, June (1980).
29. Bucklen, W. K., 'Video Digitizing Gets Extra Bit of Resolution from Flash ADC', *Electronic Design* **29**, 75–9, January 22 (1981).
30. Solomon, J. E., 'The Monolithic Op Amp: A Tutorial Study', *IEEE Journal* **SC-9**, 314–32, December (1974).
31. Giles, J., and Seales, A., *A New High-speed Comparator: The Am 685*, Application Note, Advanced Micro Devices, June (1972).
32. Saul, P. H., Ward, P. J., and Fryers A. J., 'A 5 ns Monolithic D/A Subsystem', *1980 ISSCC Digest of Technical Papers* 18–19.
33. Slemmer, W. C., 'High-speed Low-power Strobed Comparator', *IEEE Journal* **SC-5**, 215–20, October (1970).
34. Kuijk, K. E., 'A Fast Integrated Comparator', *IEEE Journal* **SC-8**, 458–62, December (1973).
35. Dendinger, S., 'Try the Sampling Comparator in Your Next A/D Interface Design', *EDN* **21**, 91–5, September 20 (1976).
36. Stein, K. U., et al., 'Storage Array and Sense/refresh Circuit for Single-transistor Memory Cells', *IEEE Journal* **SC-7**, 336–40, October (1972).
37. Yee, Y. S., Terman, L. M., and Heller, L. G., 'A 1 mV MOS Comparator', *IEEE Journal* **SC-13**, 294–7, June (1978).
38. Yukawa, A., 'A Highly Sensitive Strobed Comparator', *IEEE Journal* **SC-16**, 109–13, April (1981).
39. Naylor, J. R., 'Digital and Analog Signal Applications of Operational Amplifiers, Part II: Sample and Hold Modules, Peak Detectors adn Comparators', *IEEE Spectrum* **8**, 38–46, June (1971).
40. Bickers, L., 'Schmitt Trigger Circuit with Picosecond Risetimes', *Electronic Letters* **17**, 695–7, September (1981).
41. Bosch, B. G., 'Gigabit Electronics—A Review', *Proceedings of the IEEE* **67**, 340–379, March (1979).
42. Klein, M., Herrell, D. J., and Davidson, A., 'Sub-100 ps Experimental Josephson Interferometer Logic', *1978 ISSCC Digest of Technical Papers*, 62–3, 266.
43. Hamilton, C., et al., 'A/D Conversion using Superconducting Quantum Interference', presented at the Workshop on High Speed A/D Conversion 1980, 11–12 February, Portland, Oregon.

44. Tietze, U., and Schenk, Ch., *Advanced Electronic Circuits*, Springer Verlag, Berlin/Heidelberg/New York (1978), Chapter 10.
45. Widlar, R. J., 'New Developments in IC-voltage Regulators', *IEEE Journal* **SC-6**, 2–7, February (1971).
46. Gray, P. R., and Meyer, R. G., *Analysis and Design of Analog Integrated Circuits*, Wiley, New York (1977).

CHAPTER 6

Testing Converters

Some specifications describing the performance of converters have been mentioned in previous chapters. The purpose of this chapter is to outline the techniques for measuring these performance parameters. Since D/A converters are often basic building blocks of A/D converters and are needed to recover their digital outputs, testing of D/A converters will be treated first and measurement techniques for A/D converters then follow. Basic parameters used in this chapter are summarized and listed in the glossary of terms in the Appendix.

6.1 Introduction

Measurements on converters are used to check their appropriate function during development of a new device or to compare the performance of different converters. There is also a need to examine or to complete the specifications given in data sheets by the manufacturers. This is the case especially if an ADC is to be operated under dynamic conditions.[1] The parameters of interest are often related to special applications which cannot be treated here, but references will be given.[22,23]

If a converter has to be tested, a well-specified signal is applied to its input, and the output of the device under test is observed. Depending on the character of the input signal and the way in which the resulting output signal is analysed, the measurement techniques can be separated into qualitative and quantitative methods and, further, into static and dynamic operating conditions. Qualitative testing can often be done by relatively simple means, and is a quick indication of device performance. One of the difficulties associated with quantitative testing is the large number of samples to be handled; for example, a 12-bit D/A converter has 4096 different input and output signals. Sometimes the effort can be reduced if the conversion principle of a device is known and a special testing strategy can be found. As will be seen below, it is advantageous to describe the real performance of a converter by its input/output relationship under given testing conditions or by the errors, i.e. the deviations of the actual transfer characteristic from the ideal one. This gives more information on possible error sources in a converter than overall specifications, such as signal-to-noise ratio.

6.2 Measurements on Digital-to-Analog Converters

In an analogy to the formulae given in section 1.4 a unipolar analog output signal

S_a of a DAC using straight binary code can be written

$$S_a = \frac{S_{ref}}{2^n} \cdot (b_{n-1} \cdot 2^{n-1} + b_{n-2} \cdot 2^{n-2} + \ldots + b_0 \cdot 2^0)$$

where S_{ref} is the nominal full-scale output FS and $S_{ref}/2^n = Q$ is the nominal size of one quantizing interval (the corresponding quantities in normalized form are s_a and q; see section 1.4). The output shows $m = 2^n$ different levels and the maximum is $S_{amax} = Q \cdot (m - 1)$. Usually, S_a is a voltage or a current.

The input/output relationship of an ideal DAC described by the formula above is drawn in Fig. 6.1(a). If, for example, the actual output voltage of a DAC under test is compared with the theoretical output referred to a standard voltage source, one gets a measure of absolute accuracy of the device under test. From this point of view, the characteristics of Fig. 6.1(b) show some errors. In many cases, it is sufficient to consider relative accuracy. In the example of Fig. 6.1(b) the lower staircase has a perfect relative accuracy because there is no offset, and each increase in the analog output—due to the 1 LSB increase of the input codeword—has the same size q' and there is a constant slope over the full range. If the output signal were amplified, the so-called 'gain or full-scale error' could be nulled out and an absolutely perfect characteristic achieved. The dashed staircase exhibits an 'offset error'. Apart from the fact that this 'error' could also be nulled out, an offset of one half the quantizing interval is sometimes required in order to represent the 'lowest' quantizing interval of an ADC by its centre level (see Fig. 1.2). Therefore it may be more interesting to know the adjustment range of gain and offset, long-term stability, and temperature drift of a converter rather than its absolute accuracy.

Figure 6.2 shows the real transfer characteristic of a DAC that also exhibits relative errors. First, the analog output does not monotonically increase for increasing digital inputs. Second, the quantizing intervals show different heights q_k. This error is commonly described by the differential non-linearity $dq_k = (q_k - q)/q = (Q_k - Q)/Q$, where $q = 1$ or Q is the theoretical size of one quantizing

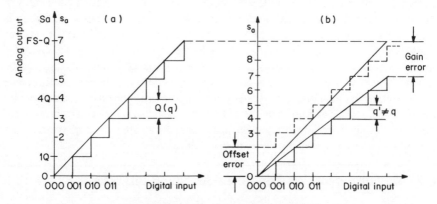

Fig. 6.1 Input/output relationship of a 3-bit, unipolar digital-to-analog converter. (a) Absolute perfect; (b) relative perfect

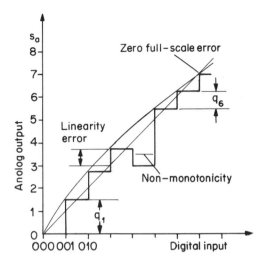

Fig. 6.2 Transfer characteristic of an actual 3-bit digital-to-analog converter (for the theoretical characteristic see Fig. 6.1(a))

interval. In the case of relative accuracy it is $Q = (S_{amax} - S_{amin})/(m - 1)$, where the output signals S_{amax} and S_{amin}, respectively, correspond to an input with all bits 'high' or 'low'. Frequently, the differential non-linearity is also given as a percentage. Considering the whole characteristic, local errors dq_k accumulate to the so-called integral non-linearity error. This is shown in Fig. 6.2 as the difference between the curve connecting all corners of the stairs and the reference line connecting zero and the corner at the maximum level. Sometimes a 'best straight line' is used as a reference yielding better specifications in the data sheet but not a better DAC. More comments on specifications and errors can be found in references 2–4.

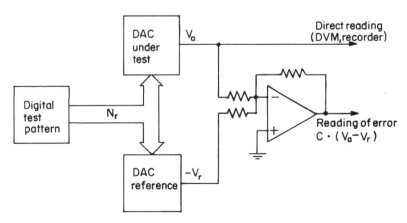

Fig. 6.3 Basic structure for testing digital-to-analog converters

The basic structure of a test set-up is given in Fig. 6.3. A digital test pattern is generated and drives the input of the converter under test. The output signal—for example, a voltage V_a—is directly observed on a voltmeter and related to the digital number N_r at the input in order to detect errors. If a DAC of high accuracy is provided as a reference whose input is fed in parallel with the sample and whose output signal V_r is subtracted from the analog signal V_a, the error can be read immediately. Modifications of this measuring set-up will be demonstrated in the following sections.

6.2.1 Static Testing of Digital-to-Analog Converters

In the case of static testing the digital input signal of the device under test is constant for a long time compared with the settling and response times of the converter and the auxiliary circuits.

In a simple version, the pattern generator of Fig. 6.3 is a manually operated bank of switches that link the inputs of the DAC with digital 'high' or 'low' levels. All possible 2^n code words must be generated and all corresponding output levels measured. In this way, monotonicity will be checked and the quantizing intervals Q_k—the output signal change for a 1-LSB change at the input—can be determined and differential and integral non-linearity calculated. Of course, careful zero and gain adjustments have to take place before testing because they affect the results. For example, in Fig. 6.2, gain is adjusted for a zero full-scale error. The peak linearity error is approximately $+0.75 \cdot q$. If gain would be somewhat diminished, there would be an error at full scale (for example, approximately $-0.5 \cdot q$), but the peak linearity errors would be 'minimized' (approximately $\pm 0.5 \cdot q$).[3] If the weight of each bit is completely independent of the on/off state of the other bits, the effort for testing can be reduced. It is sufficient to measure the analog output corresponding to each bit and to calculate resulting errors by superposition of the individual errors.[5] In so doing, zero and gain errors have to be taken into account if they were not nulled out previously. Low superposition errors can be assumed for DACs of the R–2R–ladder type but not for devices such as the voltage divider type, for example (section 4.4).

To assess quickly the characteristics of a DAC, dynamic programming of the inputs to the device is used.[3] In the 'bit-scan' mode, the bit-inputs to the device under test and the reference DAC in parallel are turned on sequentially, and the DAC error is displayed on an oscilloscope or a strip-chart recorder. Commonly, two additional 'time slots' are provided for the status 'all bits on' and 'all bits off' to facilitate full-scale and zero calibration. In the so-called 'count-mode', the pattern generator is a binary counter, so that all possible 2^n code combinations are generated and the analog output signals V_a and V_r exhibit a staircase waveform. The error display is recorded on a strip-chart and is ideally a straight line if correct calibration of the set-up has been previously carried out.

Another technique to check differential non-linearity over the whole range is achieved by again driving the DACs from a counter but with a digital subtractor inserted in front of the reference DAC. So it is offset by one digital count (the

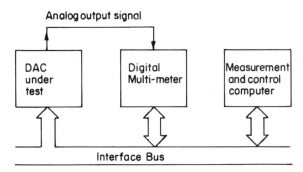

Fig. 6.4 Automatic testing of a DAC

digital number at its input is $N_r - 1$) and the error record theoretically reads a value corresponding to one LSB. For all DACs of the binary current weighting structure, measurements at the major carry transitions are of special interest. At these transitions, a carry occurs and one or more bits change if one LSB is added. In the case of 4 bits, these critical transitions are 0001/0010, 0011/0100, 0101/0110, 0111/1000, 1001/1010, 1011/1100, and 1101/1110.

Especially for evaluation of dynamic response characteristics it is convenient to cycle the DAC input periodically through a few counts on either side of interesting code transitions. For this purpose, the test pattern has to be a digital dither signal.[3]

Though high-accuracy reference DACs with resolutions up to 20 bits are reported,[6] in most cases, direct reading of the output signal S_a by a digital multimeter (DMM) is preferred. If it has a digital output, all specifications of interest can be determined by digital means without need of any further analog processing. Due to the large quantity of data, and especially for measurements in production, automatic testing is desirable.[35]

Figure 6.4 shows a test circuit where all digital data are generated and handled by a computer or microcomputer connected to the DAC under test and the DMM via an interface bus.

6.2.2 Dynamic Testing of Digital-to-Analog Converters

When a digital-to-analog converter is updated at its input with high speed relative to its internal response times, accuracy decreases. However, the transfer characteristic cannot be measured for high-speed operation. Instead of the input/output relationships, single parameters such as settling time, glitch 'energy', and overall performance such as signal-to-noise ratio have to be determined.

Settling time is defined as the amount of time between the change of an input number and the settling of the output signal to within a certain error band around the final value. Commonly, the width of this error band is $\pm Q/2$, as shown in Fig.

6.5. For a given response, settling time increases with the number of bits. Settling time includes bandwidth and slew rate limitations, transients, and pickup disturbances, and sometimes delays are also included. Because of slew rate limitations and non-linear effects (see section 5.3.2), full-scale settling time may be a worst-case specification compared with settling time for input changes of only 1 LSB in a bit-to-bit tracking mode. But if glitches occur at major transitions, settling time will also increase significantly in this operating mode.

A test set-up for measurements of settling time, needed for an output change from zero to full-scale or reverse, is shown in Fig. 6.6(a). The input of the device under test is driven, with all bits in parallel, by a pulse generator supplying a square wave and its inverse function as a reference V_r. The offset of the reference and the voltage divider are adjusted in such a way that there is a zero volt error signal at the input of the scope in the steady state, i.e. when the DAC output has settled and the digital input does not change. The oscilloscope's overdrive during switching transients is limited by the clamp diodes. For calibration of the display, the LSB is switched to logic '0' or logic '1', establishing a 1 Q-band centred about the level corresponding to a full-scale or zero DAC output. After calibration, the LSB is again linked with all other bits in parallel and the settling time to $\pm Q/2$ of the final value is measured (Fig. 6.6(b)).

A modified test configuration for measuring full-scale settling time requires a high-speed comparator (Fig. 6.7,[3]). Full-scale settling time is measured as follows. First, the LSB is set to 'dynamic' and the threshold level is adjusted to bias the comparator into its linear operating region for a full-scale output of the DAC. Therefore in the steady state with all bits '1' the comparator exhibits a level at its output between logic '1' and logic '0'. Second, the LSB is switched to 'zero', reducing the maximum DAC output level by one quantizing interval Q to

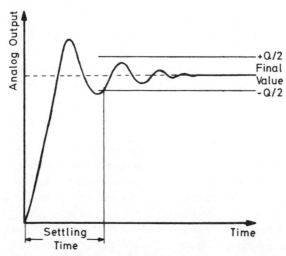

Fig. 6.5   Dynamic response of a DAC's output (pure delay omitted)

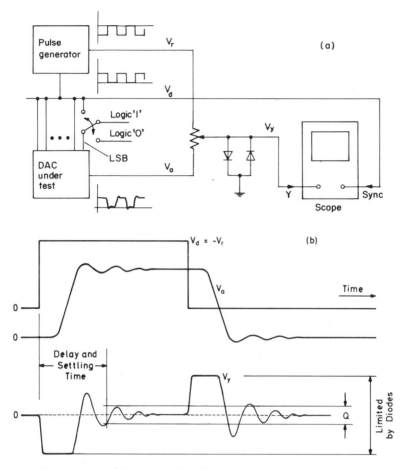

Fig. 6.6 Zero and full-scale settling time measurement. (a) Test set-up; (b) waveforms, constructed from the response in Fig. 6.5

establish a 1Q-band on the oscilloscope display. The comparator must remain in its linear region. Third, the LSB is again set to 'dynamic' and settling time is measured from the time the digital input code changes until the comparator output remains within the 1Q-error band centred about the final value. An example of waveforms during measurement is shown in Fig. 6.8. Resolution is restricted by unavoidable noise and thermal effects within the comparator.[3] Of course, high-speed, high-gain comparators tend to oscillate. Therefore the design of the measuring circuits and the connections is critical, and the set-up is restricted for laboratory use.

Transient spikes at major carry transitions may contribute significantly to settling time if, for example, the switches in a ladder network DAC show different response times (section 4.3). The worst case will be at FS/2, where all bits change (0 1111–1 0000). To measure settling time and to observe the 'glitch' at this

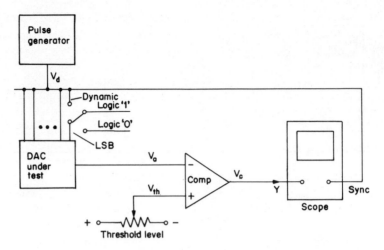

Fig. 6.7 Settling time measurement using a biased comparator

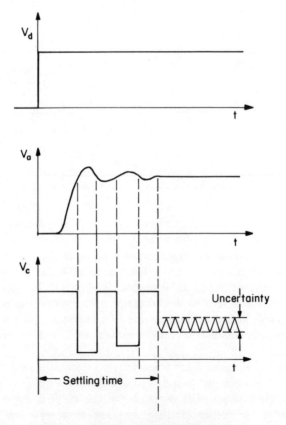

Fig. 6.8 Signal waveforms of Fig. 6.7

transition, the MSB must be driven by a signal of opposite phase to the other bits in parallel. In the case of DACs with bipolar output range, the output voltage at midscale is close to zero, so that overdriving of an amplifier or oscilloscope does not take place.

It is necessary to use a pulse generator with synchronous normal and complement pulse outputs to avoid any time skew between bits at the DAC input. Of course, this condition has to be met for all measurements if simultaneous transitions of the inputs are required.

Glitches—sometimes denoted as glitch 'charge' or 'energy'—are the areas under the plot of voltage versus time on the tube face of an oscilloscope (Fig. 6.9; waveforms are to be approximated by triangles). If glitches cannot be reduced by design of the DAC circuit itself, they are smoothed by suitable filtering or by deglitching with a sample-and-hold circuit which samples the DAC output after settling. However, in high-speed applications, the signal transients introduced by a sample-and-hold circuit are likely to be worse than those they are eliminating. For example, 8-bit DACs available for video applications show glitches in the range of 10×10^{-12} to 50×50^{-12} V $\times$ s and a 10 bit/50 MHz DAC is reported whose glitch energy in the worst case transition is 30×10^{-12} V $\times$ s.[36]

For testing other than midscale transitions, appropriate driving of the digital inputs must be chosen. Using a digital dither generator, the DAC can be periodically cycled through the transition of interest.[3]

To conclude, accurate measurement of settling time for high-resolution and high-speed DACs is associated with many practical difficulties due to bandwidth of test instruments, thermal unbalance effects, and unavoidable noise. When the output voltage of a DAC is observed on a scope, resolutions of nanoseconds and millivolts are needed. (Current-output DACs must be terminated directly in a resistor or fed to the summing node of a high-speed operational amplifier to obtain a voltage.) When measuring full-scale settling time, a large overload of the scope occurs before the output has reached a value near the final one. This difficulty also exists if a window comparator is used for detecting when the output crosses

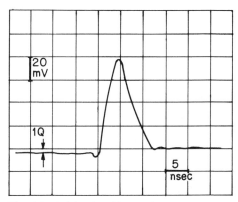

Fig. 6.9 Midscale glitch of a 8-bit video-DAC with current output (load 75 Ω)

the error bounds. Overload causes at least degradation of the response times of these indicators. Therefore very fast limiters or limiting and offsetting amplifiers usually have to be provided.[2,7]

The worst-case settling time of a DAC dictates the maximum word rate at the digital input, and this should be less than the reciprocal of the settling time. If this limit is taken into account, the dynamic overall performance of a DAC can be measured when the device under test is driven with digitally generated samples of a sinusoid.[8] The analog sinewave available at the output of the DAC after appropriate filtering can be analysed by measuring the signal-to-noise ratio and spectral purity. If the influence of the recovery filter and the sin x/x decay are taken into account (see Chapter 4), the measured values will specify the dynamic accuracy of a DAC.

6.3 Measurements on Analog-to-Digital Converters

There is a basic difference between analog-to-digital and digital-to-analog conversion. In a DAC, a digital number at its input produces a certain analog level at its output (Fig. 6.1), whereas the digital output of an ADC represents a certain signal interval, not only one level. Therefore a perfect quantizer also shows an inherent quantizing 'error' of plus or minus one half quantizing interval in the best case (Figs 1.7 and 6.12).

For testing of an ADC, the threshold levels of its quantizing characteristic must be determined. This can be performed by using an adjustable voltage source in connection with a high-resolution voltmeter to get a well-specified input to the ADC. The voltage level is increased or decreased until the code word at the output of the device under test changes, indicating that a transition between adjacent quantizing intervals occurs at a certain voltage.

In practice there will be no sharp transition between states due to the finite gain of comparators, to noise and ripple superimposed to the signal, and to the internal noise of an ADC.[9] Figure 6.10 shows the probability of the codes corresponding to the intervals Q_k and Q_{k+1} below and above a threshold level which can be defined as the point where codes C_k and C_{k+1} exhibit the same probability of 0.5. Notice that this actual transition voltage V_k is displaced from its nominal value V_n, where a sharp transition should occur. This effect described must not be confused with 'noisy transitions' of ADCs containing comparators with too large decision noise.[3]

The input/output relationship of a 3-bit ADC is given in Fig. 6.11. Absolute and relative accuracy can be defined in an analogy to DACs.[4,10] Only relative accuracy is considered in Fig. 6.11. To derive the ideal characteristic the range from zero to the transition level at the upper end is divided into $2^n - 1$ equal spaced intervals Q, yielding the theoretical staircase shape as a reference. As pointed out in section 6.2, in this case no zero errors or full-scale gain errors exist for the actual characteristic. But the threshold levels deviate from their theoretical values $V_n = k \cdot Q$ ($k = 1 \ldots (2^n - 1)$), causing different sizes of quantization

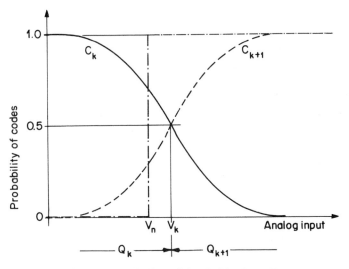

Fig. 6.10   Definition of threshold voltage V_k

intervals Q_k and a changing slope of the staircase over the full range (gain error, non-linearity).

Different quantizing intervals Q_k can also be described in terms of differential non-linearity

$$dq_k = \frac{Q_k - Q}{Q} = q_k - 1$$

If an ideal DAC is driven from the output of an ADC, the difference between the reconstructed analog output V_a and the input level V_x versus the analog input

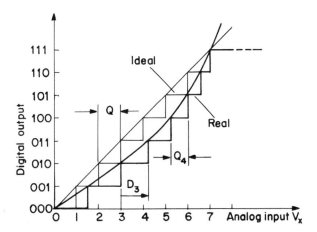

Fig. 6.11   Transfer characteristic of a 3-bit unipolar ADC

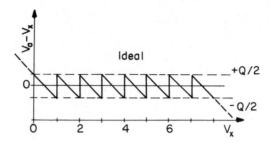

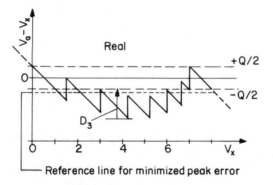

Fig. 6.12 Quantizing errors corresponding to the transfer characteristics of Fig. 6.11

exhibits the well-known sawtooth shape. In Fig. 6.12 a DAC with quantizing intervals of size Q equal to those of the ADC is presumed. Furthermore, an offset of $+Q/2$ is provided within the DAC to yield quantizing errors limited to $\pm Q/2$ for a perfect characteristic.

Any real ADC will cross these bounds. The deviations from them are frequently used for specifying (integral) non-linearity. Due to the unity slope in Fig. 6.12, the magnitudes of the peaks are the same as the displacements $D_k = V_k - k \cdot Q$ of the actual upper threshold voltages of quantizing intervals Q_k (see D_3 in Fig. 6.11). Displacements can also be calculated by accumulation of differential non-linearities

$$D_k = \sum_{j=1}^{k} dq_j$$

Usually, only maximum errors are specified. They are minimized if a suited reference line is established (Fig. 6.12). This corresponds to introducing additional offset in the DAC.

Until now, an ADC with an unipolar input range has been assumed. The theoretical characteristic for bipolar input signals is shown in Fig. 6.13 on the left.

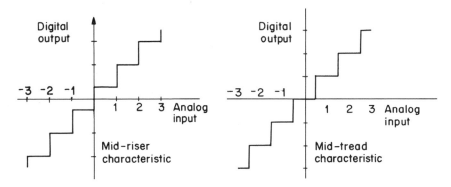

Fig. 6.13 Transfer characteristics of bipolar ADCs

This is a so-called 'midriser', because it changes state when the signal crosses zero. (A zero error or zero offset can be defined.) But to avoid idle noise for a zero input, the ADC is mostly offset by $+Q/2$, as shown in Fig. 6.13 on the right. In this case, the usable full-scale range reduces to $(2^n - 1)$ intervals. For such a midtread characteristic, a 'zero error' definition has no sense. As mentioned for DACs, in many applications the specification of adjustment range of offset and gain and parameters such as long-term stability are more useful.

If one takes into account the fact that non-linear ADCs exhibit quantizing intervals of different nominal size, performance of such devices can be specified and measured in a way similar to that for linear ADCs. Therefore only the linear case is treated in the following chapters.

6.3.1 Static Testing of Analog-to-Digital Converters

When developing a converter or making a first-order check it is often sufficient to verify the transfer characteristic in a qualitative way. Such measurements can be performed with a set-up shown in Fig. 6.14. A low-frequency triangular waveform or a slow ramp signal is fed to the analog input of the ADC and to an oscilloscope operating in the x-y mode.[11,12] The digital output signal of the device under test (DUT) is recovered by a DAC whose output voltage is connected to the y-input of the scope. Thus a transfer characteristic, as shown in Fig. 6.11, will be displayed on the tube face and errors such as non-monotonicity, large differential non-linearities, hysteresis, and noisy transitions can be indicated visually.

Because only the width of the 2^n stairs but not their heights are considered, the DAC has no influence on the test if it is monotonic. The sampling frequency must be high compared with the test signal frequency in order to ensure a code change whenever the test signal crosses a threshold voltage. Operating in the x-y mode, the displayed characteristic is independent of the linearity of the input waveform.

Of course, an x-y strip-chart recorder could also be used for more quantitative evaluation of the converter's performance. Using a DAC with a resolution of 2 to 4 bits more than the ADC, an error plot, as shown in Fig. 6.12, could also be

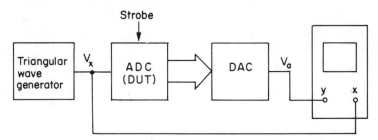

Fig. 6.14 Test set-up for measurement of the transfer characteristic

displayed with sufficient accuracy. There is no need for a DAC if the sweep output of an oscilloscope operating in its normal y-t mode is used as a test signal and the LSB's digital output is displayed on the tube face. With a binary code assumed, the LSB should exhibit a square wave. A non-constant duty cycle indicates different sizes of the quantizing intervals.

It is difficult to observe the staircase on the scope for resolutions of more than 7 or 8 bits. Therefore zooming is implemented with the 'dynamic crossplot' or 'dither pattern test' that can be explained by referring to Fig. 6.15.[3,10] A low-frequency triangular dither signal of small amplitude is superimposed on a bias voltage and fed to the input of the device under test. (The dither waveform could also be a sine wave, since the oscilloscope is operating in the x-y mode.) A window covering only a few quantizing intervals is shifted through the whole range of the ADC by adjustment of the bias voltage. The conversion rate must again be high so that the ADC will track its analog input exactly. Only the LSB and the adjacent bit are used to recover an analog signal for display. Therefore the 2-bit DAC is usually formed with two resistors having weights of 2R and R directly connected to the outputs of the two bits of interest. Because all four-level staircase waveforms look alike, in Fig. 6.15 a digital display of the complete word of the ADC output is provided for identification.

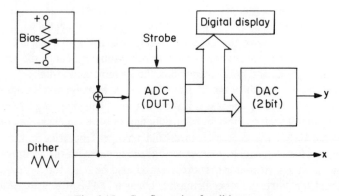

Fig. 6.15 Configuration for dither test

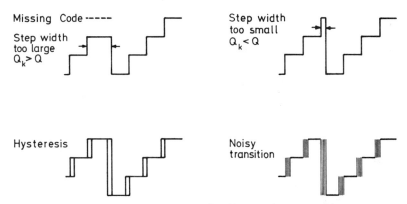

Fig. 6.16 Dither test patterns (oscilloscope in x–y mode)

Identification would also be ensured if the bias voltage is measured by a voltmeter or if it is taken from a programmable voltage source. In another proposal, the MSB is used as an additional bit for shifting the staircase waveforms on the display, yielding approximate information about the range under test.[5] Some examples of possible test patterns are shown in Fig. 6.16. With proper calibration procedures before measuring, offset and gain errors and their temperature coefficients can also be determined.[10]

Though new types of oscilloscopes containing microprocessors promise waveform measurements with better accuracy, all the principles described above appear qualitative rather than quantitative. In addition, the test set-ups are manually operated, thus needing a lot of time for measurements.

One step towards quantitative, automatic testing is to replace an analog waveform generator by a high-resolution, high-accuracy DAC driven from a binary counter. But even if the DAC exhibits ideal performance, there are inherent problems due to the quantized shape of the test signal. The digital output of the ADC under test and the digital input of the DAC cannot be compared in a simple way, and threshold voltages can only be determined with limited resolution.[13]

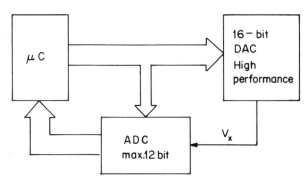

Fig. 6.17 Automatic test of static performance

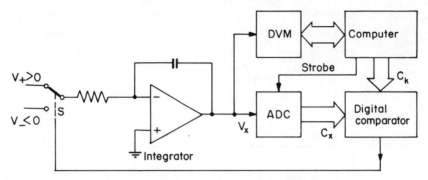

Fig. 6.18 Feedback approach to ADC testing

Burney[14] describes a system that automatically tests 12-bit A/D converters using a self-calibrating 16-bit D/A-converter as a reference, and Fig. 6.17 shows a simplified structure of the set-up. Static performance is tested under the control of a microcomputer. When signal ranges of both, ADC and DAC, are matched by automatic adjusting of offsets and gains, the code transitions of the device under test are searched, including a hysteresis check.

McClellan[15] claims to need only a DAC with a resolution of two bits more than the ADC to be tested. He estimates the transition levels from the code's transition point, i.e. the analog input level at which the digital output totally changes to that code (compare Fig. 6.10). The threshold is estimated as the midpoint between a present analog level and the preceding one. This approximation becomes better with increasing resolution of the DAC. From the data recorded by a computer, the transfer characteristic of the ADC is calculated and a variety of static performance analysis options are provided.

Another computer-aided technique is published by Havener[16] and Corcoran et al.[17] This is a feedback approach, as shown in Fig. 6.18. A commanded code C_k is produced by a computer—in a very simple configuration by a counter—and

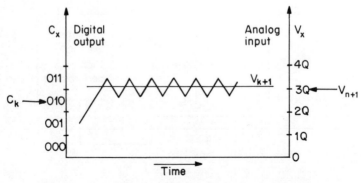

Fig. 6.19 Run of the analog input voltage V_x in Fig. 6.18

compared with the output code C_x of the ADC. If $C_x = C_k$ the input of the integrator is switched to a voltage V_- and the input signal V_x of the device under test ramps until $C_x > C_k$ and the switch changes to the V_+ position after a certain loop delay, whereupon the integrator's output ramps down until V_x produces the correct code C_k or a code $C_x < C_k$. Figure 6.19 shows a simple example when the commanded code C_k is fixed at 010 and the ADC is in continuous operation. The average value of input voltage V_x is measured by the digital voltmeter DVM and equals the threshold voltage V_{k+1} between the intervals corresponding to the commanded code word C_k and the code C_{k+1}. Averaging implies the definition of transition points, as already given in connection with Fig. 6.10.

The maximum sampling rate for the test is dictated by the DVM. The integration rate must be chosen such that the analog voltage changes by much less than one quantization interval between successive conversions due to the finite averaging time of the DVM.[17] The converter under test must be monotonic to ensure error-free operation of the feedback system. To detect non-monotonicity, it is convenient to connect a 4-bit DAC to the 4 LSBs of the ADC, thus monitoring its output on an oscilloscope. (The resulting waveforms are similar to that of Fig. 6.16.) With additional equipment, missing codes could be searched.[16] Such a feedback approach is also used for noise measurements in the calibration service of the American National Bureau of Standards, as described in reference 37.

As a result of a test, Fig. 6.20 shows an error plot for illustration. Errors were calculated from the commanded codes C_k, the corresponding nominal threshold voltages V_n, and the actual threshold levels V_k which are available in digital form

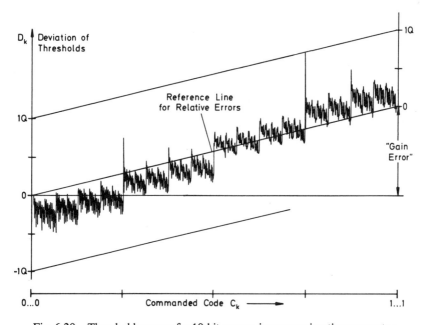

Fig. 6.20 Threshold errors of a 10-bit successive approximation converter

at the output of the DVM. Gain and offset were not adjusted before measurements, therefore a reference line is drawn so that zero errors exist at the beginning and at the end of the range. The recurrence of similar errors along the input range indicates that they are caused to a large extent by the R–2R ladder of the DAC implemented in the converter under test.

Reversing the principle of pulse height analysis used in nuclear research, the performance of A/D converters can be evaluated from the probability distribution of code words if the probability density of the analog test signal is known.[18,19] A configuration for static testing based on such an approach is shown in Fig. 6.21. The minicomputer (or an external clock generator) strobes the ADC and accepts its digital output when conversion is ready. The number of events of each of the 2^n code words is counted within the desired number z of samples, which is controlled by software.

The analog input signal is free-running with respect to the strobe. A triangular waveform is preferred for testing, since its probability density function is constant along the full-scale range V_{FS}.

In this case, each code word of an ideal ADC occurs with the same frequency $f_0 = z/2^n$ and the nominal size of each quantizing interval is $Q = V_{FS}/2^n = V_{FS} \cdot f_0/z$. A non-perfect ADC exhibits different frequencies f_k of code words C_k corresponding to the intervals Q_k. Their size is given by $Q_k = V_{FS} \cdot f_k/z$ or $q_k = Q_k/Q = f_k/f_0$. These formulae are valid if z is large enough for statistical variations to be neglected. A satisfactory statistical confidence level will be obtained with 10^6 samples and more for resolutions up to 8 bits.[21]

For example, Fig. 6.22 shows the histogram, i.e. the frequency ratio of occurrence $F_k = f_k/f_0$ plotted versus the code words C_k, of a 4-bit converter under test. Note that the values of the two code words representing the boundaries of the analog input range cannot be measured in principle, except that an ADC supplies additional information when it is overdriven. But this means that the ADC has to provide more than n bits, i.e. exactly $\log_2(2^n + 2)$ bits. It must be ensured that strobe pulses are a pseudo-random binary sequence to prevent beat frequencies affecting the measurements. This would be shown by a periodical structure of the histogram.

If the sizes of the quantizing intervals Q_k are calculated by the formulae given

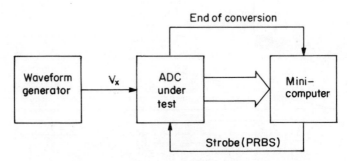

Fig. 6.21   Test set-up for statistical evaluation

above, the transfer characteristic of a converter can be constructed as shown in Fig. 6.23 (intervals Q_1 and Q_{16}, represented by 0000 and 1111, are substituted by their nominal values). But it is not necessary to draw this input/output relationship. The frequency distribution (the histogram) will deliver equivalent information, because the length of a certain bar, representing a certain frequency

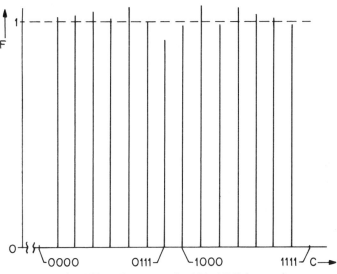

Fig. 6.22   Histogram of a 4-bit ADC (see text)

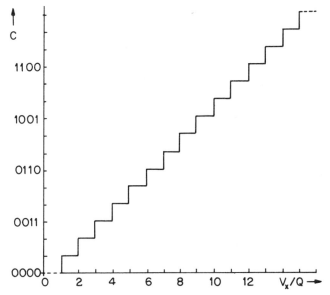

Fig. 6.23   Transfer characteristic corresponding to Fig. 6.22

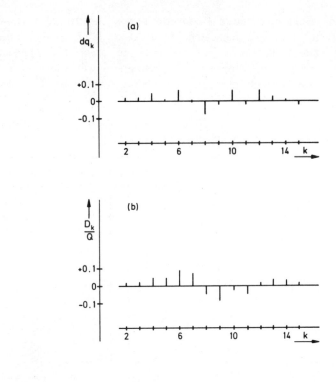

Fig. 6.24 Static performance specifications of a 4-bit ADC. (a) Differential non-linearities; (b) displacements of thresholds; (c) error plot

ratio F_k, is reciprocal to the slope of the transfer characteristic within an interval Q_k.

Knowing quantizing intervals Q_k or frequencies f_k, differential non-linearities dq_k, as defined at the beginning of section 6.3, can be determined. With $dq_k = f_k/f_0 - 1 = F_k - 1$, they show themselves in the histogram supplemented with a horizontal line $F = 1$ (a dashed line in Fig. 6.22). Displacements D_k of threshold levels can be calculated as well as the associated error plots (see Figs 6.11 and 6.12) and the resulting signal-to-noise ratio (see section 6.3.2), available at the output of a hypothetical perfect DAC linked to the device under test.[20,21] Examples corresponding to Figs 6.22 or 6.23 are shown in Fig. 6.24. Further details of this method of measurement are given at the end of the next section.

6.3.2 Dynamic Testing of Analog-to-Digital Converters

The test set-up (Fig. 6.14) which is used for static qualitative measurements can also be used for dynamic tests. The maximum throughput rate of an ADC will be determined if the sampling frequency is increased until the transfer characteristic V_a versus V_x or the reconstructed low-frequency signal $V_a(t)$ exhibits errors such as non-monotonicities or large differential non-linearities. At the upper limit of the conversion rate of an ADC there is usually a sharp breakdown in performance, seen in the input/output relationship displayed on the oscilloscope.

Smaller errors are likely to be associated with the strobe frequency. For example, a pulse feedthrough or a hold-droop of an internal sample-and-hold circuit will cause some change in the performance of an ADC, depending on sampling frequency. To detect such errors, the quantitative methods of measurements, as already described in the previous section, can be used as insofar as they allow variation of the sampling frequency. Otherwise, pure dynamic techniques, as explained later, must be used.

Conversion time of an A–D converter is defined as the time elapsing between actual sampling of the analog input signal and the occurrence of the corresponding digital output. Conversion time need not to be the inverse of the maximum word rate of an ADC. (As mentioned in section 3.3.6, a pipelining configuration yields a higher throughput rate.) Therefore the conversion time must be measured separately. This can be done with a set-up similar to that of Fig. 6.14 in the following way. A triangular waveform V_x is locked to a pulse generator strobing the digitizer at twice the signal frequency, i.e. $f_s = 2f_x$. In Fig. 6.25, for simplicity the sampling pulses are assumed to have a zero pulse width and the response times of the DAC are neglected with respect to the conversion time of the ADC and the sampling period. The reconstructed analog output signal changes between two levels. (The corresponding bits at the converter output could be observed instead of the analog output signal.) Two cases are distinguished in Fig. 6.25:

(1) The internal conversion time $t_c \approx t_{c1}$ is smaller than the sampling period t_s and the digital output of the converter under test changes when conversion is ready. Internal conversion time is of special interest to the designer and will be measured by him during, for example, development.
(2) The digital output of an ADC is usually latched. Therefore data do not change state until a new sampling pulse occurs and a time $t_{c2} > t_{c1}$ is measured.

In practice, the internal conversion time may be longer than the minimum time between strobe pulses, as described for pipelining converters (section 3.3.6), and there will be a further delay between the sampling point and the digital output due to processing of analog and digital signals inside the converter, for instance, intermediate storage and delay. An example of a video-digitizer is given in Fig. 6.26. The sampling pulse, generated internally, is fed to a sample-and-hold provided but it is not accessible to the user. Considering the ADC as a black box, the user can determine only the 'external' conversion time t_{c3}.

Tracking frequency, sometimes given as a specification, is the maximum frequency of an input signal which the ADC can track. This is ensured as long as the

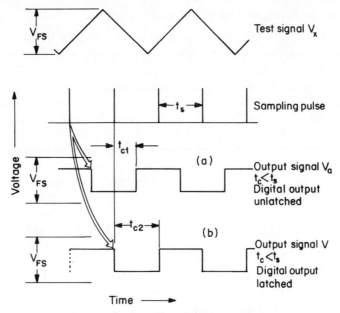

Fig. 6.25 Measuring conversion time

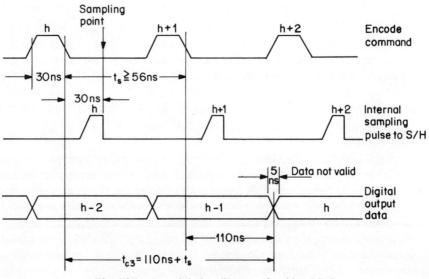

Fig. 6.26 Actual timing diagram of a video ADC

change between succeeding samples is not more than one quantizing interval Q. Tracking frequency can be derived from the maximum word rate f_{smax} of a converter.

Feeding a full-scale sine wave to the input of a converter, the tracking frequency is $f_{smax}/(\pi \cdot 2^n)$. In the special case of a feedback converter, implementing the counting mode—the true 'tracking converter'—the tracking frequency is calculated with the word rate in the formula above substituted by the clock frequency steering the conversion steps. Of course, tracking frequency for a given signal waveform can also be measured if the recovered signal $V_a(t)$ is observed on a scope.

Analog bandwidth of an ADC can be defined and determined by methods similar to analog techniques. Using the set-up of Fig. 6.14, the input signal is a sine wave covering the full-scale range and the oscilloscope operates in the $y(t)$ mode. For detection of −3 dB bandwidth, the frequency is increased until the reconstructed output voltage $V_a(t)$ exhibits a 3 dB decrease in amplitude. To eliminate the influence of the DAC and for ease of observation, it is appropriate to examine the digital output itself by the means of a logic analyser. This must be triggered from the code words which correspond to a −3 dB output. When the signal frequency is increased to just below triggering breakdown will occur and the corresponding bandwidth limit is measured. Full-power bandwidth of an ADC can be defined as the frequency value where the 'all zero' and 'all one' code words representing the peak levels begin to diminish. This can be measured by means of a logic analyser or by observation of the recovered analog signal $V_a(t)$ in an analogy to the measurement of the −3 dB bandwidth described above.

The analog part in front of the encoder itself determines analog bandwidth. Therefore bandwidth can be measured with well-known analog techniques if the input of the encoder is accessible.

Sometimes overall dynamic performance of an ADC–DAC pair connected back-to-back is specified. For this purpose, the signal-to-noise ratio or the spectrum of the reconstructed signal are measured.[22] Both parameters indicate ac-linearity of the devices. For video applications, such tests are in use, feeding modulated ramp and staircase signals or modulated pulses to the input of the coder. The results are described in terms of 'differential gain and phase' or the pulse responses representing short-time waveform distortions.[22,23] It must be pointed out that such tests, using a back-to-back configuration, include the performance of the DAC and of the recovery filter. The errors of the ADC itself cannot be identified but only estimated.

The concept of specifying the performance of an ADC by means of its transfer characteristic is also viable under dynamic conditions if suitable care is taken. Dynamic errors of test equipment must be excluded from influencing the results.

A technique for making sure that the ADC is subjected to a high slew rate test signal whereas the scope with its poor x-bandwidth has sufficient time to respond is implemented in the test set-up of Fig. 6.27.[21] A low-frequency substitute V_r is used for steering the x-channel instead of the high-frequency input signal V_x. The condition that has to be met in every sampling point is $V_r = V_x$. The input

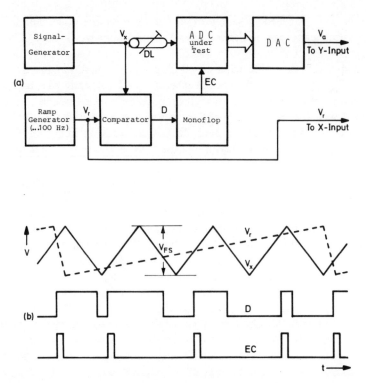

Fig. 6.27 Measurement of dynamic transfer characteristics. (a) Test set-up; (b) signals

waveform V_x and a slow ramp V_r are fed to a fast comparator which changes its output state whenever the signal V_x crosses the reference level V_r. Controlled by the user, the monoflop, following the comparator, generates an encode command EC at each low/high or high/low transition. The time needed in this path must be compensated in the signal path by an adjustable delay line DL. The sampling rate is approximately the same as the input frequency. The digital output of the device under test is recovered by a fast DAC whose output is fed to the vertical input of an oscilloscope. The scope is blanked when the reference voltage V_r ramps down.

Figure 6.28 shows the actual transfer characteristic of a 6-bit/100 MHz A/D converter digitizing a full-scale sine wave of 10 MHz. Note the considerable differential non-linearities, especially in the middle of the range where a sinusoid exhibits its maximum slew rate. For higher resolution, a larger display should be used or zooming must be provided.

Qualitative testing of ADC dynamic performance can also be done by using a beat frequency test and its related modifications. As shown in Fig. 6.29, a sine wave equal to the input range is offset in frequency from the sample rate f_s by a small amount Δf chosen so that a 1Q-change occurs between every sample at the maximum slew rate of the analog signal.[24] After reconstruction of the signal with

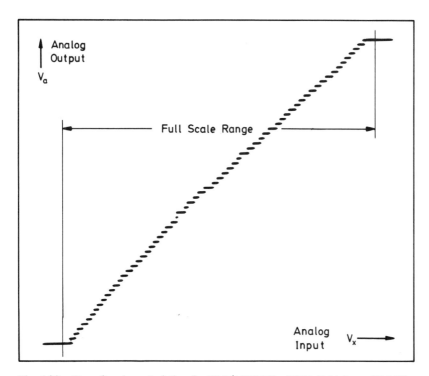

Fig. 6.28   Transfer characteristic of a 6-bit/100 MHz ADC digitizing a 10-MHz sine wave

a DAC the quantized waveform contains the low-beat frequency Δf as the fundamental that can be extracted by appropriate filtering (Fig. 6.29(a), (b)). A low-frequency output signal can be observed and analysed in a more convenient way than high-frequency waveforms and the requirements of settling time of the DAC are considerably relaxed. Bandwidth can also be measured with this beat frequency technique.

If the sine wave is slightly offset from one-half the sampling frequency, the ADC is digitizing the input waveform at points which are apart somewhat more than one half cycle, so that large changes between samples also occur. Therefore this so-called 'envelope test' is a very stringent test of response times in the ADC.[24]

Another way of producing a low-frequency output signal, when an ADC is tested with high-frequency waveforms, is a sampling technique. A system published by Connolly et al. is given in Fig. 6.30.[11] The convert command is generated by a voltage-controlled oscillator (VCO). It is frequency-locked to the analog test signal and shifted in phase through a variable time delay which is driven from a 100 Hz triangle generator. Thus the sampling points are slowly walking through the input waveform, as seen in Fig. 6.27(b). A variety of conversion rates is provided by a programmable divider that is to be passed by the

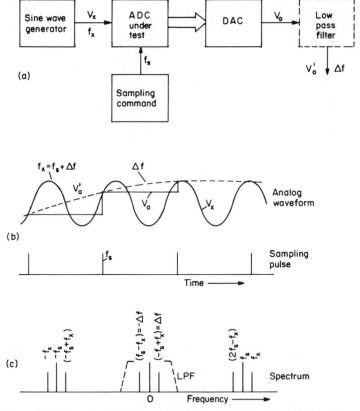

Fig. 6.29 Principle of beat frequency testing. (a) Test set-up; (b) signals; (c) spectrum

sampling pulses. Additional division of the pulse frequency by a factor of 30, for example, for strobing the latch allows the use of a relatively low-speed but high-accuracy D/A converter and scales down the frequency of the output signal. For displaying the low-frequency replicate of the input waveform, the oscilloscope operates in the x-y mode with the phase-shifting triangular wave connected to the horizontal x-input.

Until some years ago, the inaccuracies of digital-to-analog conversion and analog instrumention have been limiting factors in quantitative dynamic measurements on high-speed ADCs. The only method of adequate testing of ADCs is to feed a well-defined test signal to the analog input of the device and to derive its performance directly from its digital output, instead of using a reconstructed analog signal. Such procedures became possible by the availability of suitable digital processing systems at reasonable prices.

Fast Discrete Fourier Transform (DFT) is one of the techniques used for evaluation of ADC dynamic performance.[25,26,29] A filtered sine-wave signal is fed to the converter under test and a finite number of encoded samples appearing at a

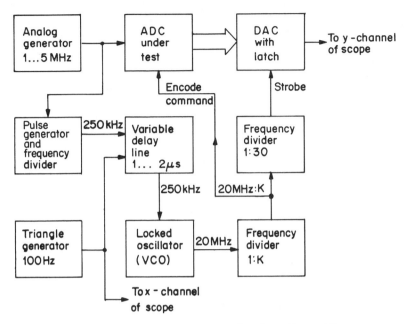

Fig. 6.30 Dynamic ADC-tester after Connolly et al.[11]

high rate f_s must be stored in a high-speed buffer memory before digital processing takes place (Fig. 6.31). The DFT converts sampled data into the frequency domain from which the linearity of the ADC transfer characteristic may be measured. In the case of a perfect converter, only the fundamental and a noise floor exist in the spectrum. The noise floor is caused by the inherent quantizing noise which is uniformly distributed over the frequency band.[28] For an n-bit A/D converter the ratio of the fundamental to the highest amplitude harmonic should be greater than $6 \cdot n$ dB (see section 2.1). Otherwise integral non-linearities of the transfer characteristic exist. The signal-to-noise ratio is computed from the DFT data by summing the magnitudes of the spectral lines of the fundamental and dividing it by the sum of the remaining spectral lines which represent noise. Due to the finite number of data samples that can be stored in the buffer, the calculated spectrum would exhibit an envelope of a sin x/x shape. Therefore successive samples must be multiplied by an appropriate weighting function to reduce leakage effects in the spectrum via the side lobes. (Pratt[27] uses $\cos^2$-weighting, Polge et al.[26] employ an approximation of Van der Maas weighting). To give an example, a 10-bit/20 MHz A/D converter is reported to exhibit a peak noise of 59.3 sB below the fundamental at a signal frequency $f_x = 0.95$ MHz and 59.6 dB below the fundamental at $f_x = 9.85$ MHz.[24]

The spectra calculated indicate the errors of an ADC but they do not directly indicate the sources of errors. Therefore Polge et al. simulate certain types of dynamic errors (gain and linearity errors and time jitter, for example) and determine the associated discrete power spectrum.[26] Comparing such a 'reference'

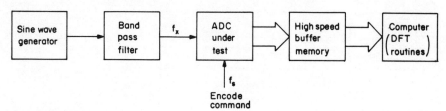

Fig. 6.31 Block diagram of a test set-up implementing Discrete Fourier Transformation

spectrum with that of the device under test, the existence of certain types of errors can be determined.

It must be pointed out that the sampling frequency depends on the signal frequency. A word rate of at least six times the input frequency is needed to investigate the presence of the third harmonic which is important for typical non-linearities.[26] Another restriction of this technique is the storage capacity of the buffer memory limiting the number of samples. For example, only 1024 samples are used in a 10-bit test set-up operating at sampling frequencies up to 15 MHz.[27] (Some software for Fast Fourier Transform Testing is given in reference 24.)

A time domain approach is used by Ochs for analysing digitizer performance.[29] The set-up for this sine-wave fitting test is given in Fig. 6.32(a). A filtered sine wave $X(t)$ is used as a test signal, represented by $x(t) = a \sin(2\pi f_x t + \phi) + b$, where a, ϕ, and b are the amplitude, phase, and d.c. offset parameters. The digitized samples at the ADC output and their analog representation are termed $X_D(kT_s)$, where k is the sample index and T_s the sampling interval. From the set of sampled data stored in the memory a reference signal $X_F(t) = a \sin(2\pi \hat{f}_x t + \hat{\phi}) + \hat{b}$

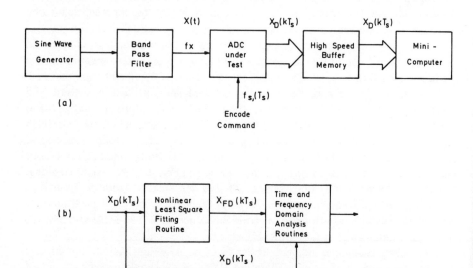

Fig. 6.32 Configuration of a time domain approach for dynamic testing of ADCs. (a) Test set-up; (b) digital processing

is estimated by a non-linear least-square fitting routine. $X_F(t)$ is taken to be the analog input to the device under test, so that $X_F(kT_s)$ and $X_{FD}(kT_s)$, respectively, are the corresponding analog and quantized samples (Fig. 6.32(b)). $X_{FD}(kT_s)$ represents the reconstruction of the output of a perfect ADC of the same resolution as the device under test. Actual and reference data are compared for error analysis, and this can be done either in the time or frequency domain.

The analog error signal is $E(kT_s) = X_D(kT_s) - X_F(kT_s)$. This will be greater than the quantizing error $N(kT_s) = X_{FD}(kT_s) - X_F(kT_s)$ of an ideal ADC. The ratio of the r.m.s. values of both is a direct measure of a converter's performance and can be expressed in terms of effective bits $n' = n - \log_2(E_{rms}/N_{rms})$. (The r.m.s. quantizing error may also be expressed as $N_{rms} = Q/\sqrt{(12)}$, where $Q = 2 \cdot a/2^n$ for a full-scale input.) A plot of effective bits versus signal frequency is a convenient description of dynamic performance. In addition, calculation of worst-case and r.m.s. encoding errors $e(kT_s) = X_D(kT_s) - X_{FD}(kT_s)$ is provided. If the encoding errors at or near zero crossings of the sine wave are computed and the relationship between amplitude uncertainty and time jitter is taken into account, jitter values and their statistics can also be derived.[29] Using simulated tests, Ochs claims that satisfactory results will be obtained with 256, 512, and 1024 sample points for an 8-bit converter. Sampling rates in these simulations varied from 0.2 to 1.6 times the frequency of the sine wave. It is pointed out in reference 24 that the test frequencies should be non-harmonically and non-subharmonically related to the sample rate of the ADC. Otherwise certain codes would occur and test results would be less reproducible. (A reproducibility to better than 0.2 effective bits is claimed in reference 24 for a consecutive number of computations for testing a 10-bit ADC.)

It must be noted that this approach cannot detect converters' gain, offset, or timing compression/expansion errors. These errors must be checked separately, the first by using static techniques, the last by comparison of the fitted frequency $\hat{f}_x$ with the input frequency f_x measured.

Of course, the analog input signal should be full scale in order to examine the complete range of the device under test. Further details about this method of measurement and some software can be found in references 24 and 30.

Another technique, proposed by Gardener et al. as an aperture performance measurement, appears to be a combination of the beat frequency and the sine-wave fitting test.[31] As shown in Fig. 6.33, the input frequency and the sampling frequency are locked. They are selected according to $f_x = (k \cdot f_s + \Delta f)$, where k is an integer and Δf a small offset frequency. The offset Δf determines the fundamental of the waveform at the output of a sample-and-hold inside an ADC or of the analog representation of the digitized waveform if digital sampling is used in a flash-mode ADC. Thus the sampling circuitry of an ADC can be exercised by high slew rate signals, whereas the succeeding circuits have to process relatively low-frequency signals. (Testing at a high slew rate is important in order to check the aperture performance of an ADC.) For example, the encoding rate was chosen to be f_s = 5 MHz and the input frequency to be f_x = 10 MHz + 1 kHz, yielding a quasi-static beat frequency Δf = 1 kHz.

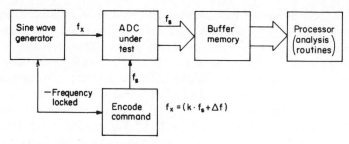

Fig. 6.33 Test set-up for aperture performance measurement

The buffer memory shown in Fig. 6.33 must accept the number of digital samples required from the output of the device under test at a rate f_s. Afterwards data are available for further processing. If the sampling rates used do not exceed frequencies around 5 MHz, a dedicated bit-slice processor system could be implemented so that a stand-alone buffer memory could be omitted.

The stored data from the ADC are compared with a synthesized ideal response generated in the processor. For comparison of static and dynamic performance, first a quasi-static low signal frequency Δf could be applied to the ADC input to produce and store reference data in the processor. As an alternative, least-square sine wave fitting routines could also be applied to the incoming digitized samples to establish a reference sine wave in the data domain.

As in the sine-wave fitting test previously described, the actual errors, their r.m.s.-values, and the effective number of bits are calculated. The use of Fourier Transform techniques is proposed to determine the level of the fundamental in the error waveform in order to cancel phase lag effects degrading the computed effective number of bits and in order to investigate the phase and amplitude responses of the sampling circuitry.[31] If sine-wave fitting is used, as mentioned above, the effects of phase lag are removed before calculation.

The histogram test introduced for static measurements is also profitably used for dynamic testing.[20] An appropriate test set-up, as an extension of Fig. 6.21, is shown in Fig. 6.34. Again, the test signal is free-running with respect to the sampling command. Thus the levels of the samples are varying as under real operating

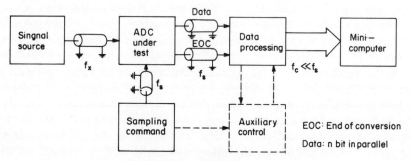

Fig. 6.34 Set-up for dynamic histogram testing[20]

conditions and all codes will be examined if the input waveform has an appropriate amplitude and a sufficient number of samples is taken. The signal frequency and the conversion rate may be chosen to be fully independent in this approach.

If a strobing technique is used to overcome the large discrepancy between the output word rates f_s of a high-speed ADC and the input rate f_c of a minicomputer, there is no need for a buffer memory. A further advantage of scaling down the data rate is that the disk memory of the computer can be used instead of the core store. Thus the desired number of samples which are controlled by software is practically unlimited. Strobing is implemented in the interface labelled data-processing in Fig. 6.34. This contains a programmable divider which generates a strobe pulse (rate f_c) whenever the computer can accept new, valid data at its input. A pseudo-random sampling command is used to avoid periodical structures in the histograms due to the sampling and strobing procedures.

For static testing, the input signal was assumed to exhibit an ideal triangular shape. However, by using waveform generators, this assumption is difficult to meet in the megahertz range. In order to overcome these difficulties, trapezoidal waveforms can be used. Due to their linear ramps, these also show a uniform distribution of all levels, except the peak levels (top and bottom). The latter must be suppressed in the data domain by hardware or software in order to avoid an overflow on the left and right ends of the histograms. For a first iteration, a pulse generator with adjustable rise and fall times may be used as signal source. To maintain a sufficient linearity of the ramps, a special generator should be implemented.[21]

Dynamic performance of ADCs depends mainly on the slew rate of the analog input signal. For comparison with Figs 6.22 to 6.24, which show the performance of an ADC at a slew rate of 15 V/μs, the results are given in Figs 6.35 to 6.36 for the same ADC tested at a slew rate of 500 V/μs.

At this high slew rate the converter under test exhibits a frequency of occurrence ratios F_k and quantizing intervals Q_k (q_k), respectively, which differ substantially from their nominal values (Fig. 6.35). The associated differential non-linearities $dq_k = F_k - 1$ can be read from the histogram, as described in section 6.31, and they are drawn separately in Fig. 6.36(a). Due to the different sizes Q_k, their threshold levels V_{k-1} and V_k are displaced from their nominal values $(k-1) \cdot Q$ and $k \cdot Q$, respectively. Related to one quantum Q and expressed in terms of frequency ratios F_k measured, these displacements d_k are[21]

$$d_k = D_k/Q = \sum_1^k F_j - k = \sum_1^k dq_j; [k = 1 \ldots (M-1)]$$

where M is the number of possible code words. If the total range of an ADC is examined, it is $M = m = 2^n$. The displacements and the resulting error plot are shown in Figs 6.36(b) and (c).

These irregularities cause additional noise, enlarging the theoretical quantizing noise $P_n = Q^2/12R$ (section 2.1), and there will be a deterioration $-(\Delta S/N)$ of the signal-to-noise ratio if an ideal D/A converter with an appropriate full-scale range

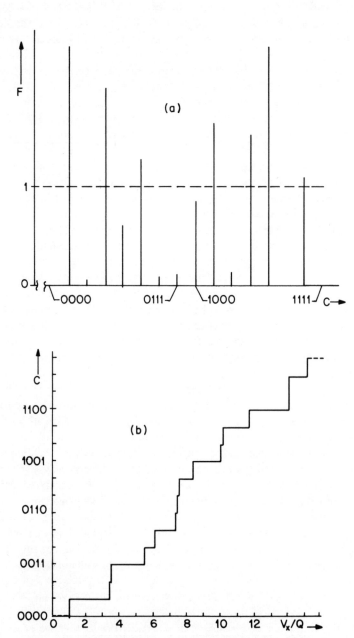

Fig. 6.35 Dynamic performance of a 4-bit ADC tested at 500 V/μs. (a) Histogram; (b) transfer characteristic

is connected to the output of the device under test.[32] The additional noise can be minimized by introducing an offset

$$V_0 = \frac{1}{M} \sum_1^{M-1} D_k \text{ or } V_0 = \frac{1}{M} \sum_1^{M-1} d_k$$

which means a shift of the dashed reference line shown in Fig. 6.36(c). In this case, the calculation of $\Delta S/N$ yields[21]

$$(\Delta S/N)_{\text{dB}} = -10 \log_{10} \left[1 + \frac{12}{M} \cdot \sum_1^{M-1} (d_k - V_0)^2 \right]$$

For a first approximation, $\Delta S/N$ can be calculated by the differential non-linearities dq_k, respectively, by the variance σ_F^2 of the associated frequency ratios F_k:

$$(\Delta S/N)_{\text{dB}} \approx -10 \log_{10} (1 + 12\sigma_F^2)$$

This approximation holds so much the better the more the ADC shows a performance where the errors are randomly distributed over the total range.

Of course, a decrease in the signal-to-noise ratio can also be expressed as a loss of bits or as the number of effective bits $n' = n + (\Delta S/N)_{\text{dB}} : 6$.

In Figs 6.24(a) and (b) and 6.36(a) and (b) the errors are plotted versus the number k of the interval considered. A very instructive representation of differential non-linearities dq and deviations of thresholds D is their cumulative frequency $CF(dq)$ and $CF(D)$. These are drawn in Fig. 6.37(a) and (b) for two different slew rates of the test signal at a fixed sampling rate. The run of the curves becomes more flat with increasing slew rates, thus indicating increasing errors. As shown in Fig. 6.37(b), for example, the converter exhibits no deviations of thresholds greater than $+0.08 \cdot Q$ at a 15 V/µs slew rate, whereas at 500 V/µs about 38% of the deviations—often specified as 'linearity', as already mentioned—exceed the $-Q/2$ limit and about 22% exceed the $+Q/2$ limit.

If a trapezoidal test signal approximates a sine wave (Fig. 6.38), the results can be associated directly with signal frequencies. For illustration, the frequency response of peak and mean values of differential non-linearities and the change of signal-to-noise ratio is given in Fig. 6.39 for an all-parallel 6-bit/20 MHz A/D converter. The experimental results for different types of A/D converters are reported in reference 21.

For testing ADCs at signal frequencies higher than 15 MHz and/or at resolutions of more than 8 to 9 bit, it will be difficult to produce a trapezoidal signal with sufficient linearity. In this case, a pure sine wave generated in an oscillator must be used. Then a certain code word occurs with a frequency

$$f_k = \frac{z}{\pi} \left(\arcsin \frac{V_k}{\hat{V}_2} - \arcsin \frac{V_{k-1}}{\hat{V}} \right)$$

where $z =$ the total number of samples, $\hat{V} =$ the amplitude of the sine wave, and V_k, V_{k-1} upper and lower threshold level of interval Q_k. An example is given in Fig. 6.40.

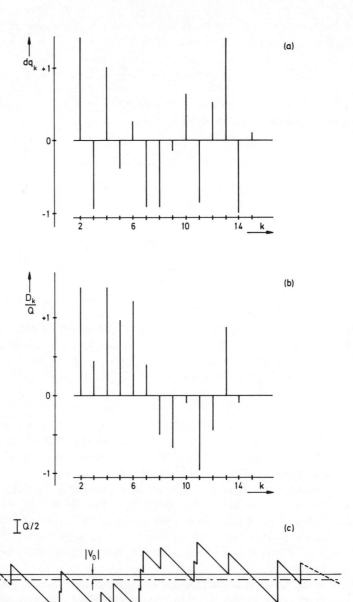

Fig. 6.36 Errors derived from the histogram Fig. 6.35(a). (a) Differential non-linearities; (b) displacements of thresholds; (c) error plot ($\Delta S/N = -9.2$ dB)

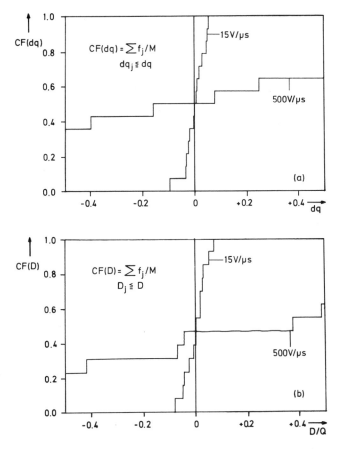

Fig. 6.37 Cumulative frequency of errors depending on slew rate. (a) Differential non-linearity dq; (b) deviation of threshold levels D

Before calculating performance, parameters such as a histogram or the corresponding numbers f_k have to be equalized. It must be pointed out that is cannot be done by a fixed scheme unless the ADC has an ideal performance. It is important that the signal range and the input range of the device under test are well matched before measurement, and the amplitude $\hat{V}$ must be known exactly. In this case, the transition voltages of each interval can be calculated, for example, when starting calculation at the upper end of the range ($V_{FS} = \hat{V}$ for a bipolar range) and each interval Q_k is bounded by the upper level V_k ($V_k = \hat{V}$ for $k = m$; bipolar range) and by the lower level[21]

$$\frac{V_{k-1}}{\hat{V}} = \sin\left(\arcsin\frac{V_k}{\hat{V}} - \frac{f_k}{z} \cdot \pi\right)$$

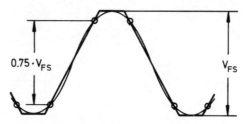

Fig. 6.38 Approximation of a sine wave by a trapezoidal wave of frequency $f_{tr} = f_x$

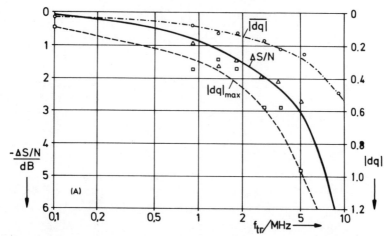

Fig. 6.39 Frequency response of some performance parameters of a 6-bit/20 MHz A/D converter (peak values $|dq|_{max}$ and mean values $\overline{|dq|}$ of differential non-linearities, change $\Delta S/N$ of signal-to-noise ratio)

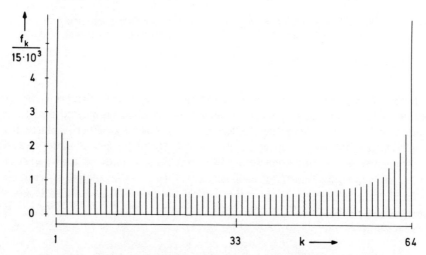

Fig. 6.40 Histogram of a 6-bit A/D converter tested with a sine wave

Thus all threshold levels and the sizes of quantizing intervals must be determined step by step. An analogous procedure must take place when a non-linear A/D converter is tested by a signal with a uniform probability density.

The internal sampling jitter (aperture jitter, see section 5.2) of an ADC can also be determined by the histogram test if the signal and the sampling command are locked together. In the case of high-speed converters, which exhibit a sampling jitter of some 10 ps, locking of two generators is not appropriate. The output of a pulse generator should be split to derive a testing signal and a sampling pulse from a single source. One of them should be delayed with respect to the other in order to adjust the analog level at the sampling point. Using only passive components for splitting and delay, additional jitter between both signals is avoided.[18]

A jitter-free ADC will exhibit only one code word if the sampled level falls within a quantizing interval, or two code words will exist due to decision noise if the analog level equals a threshold voltage.

The histogram test is a statistical approach, and the immediate relationship between the sampled analog values and the digital code words recorded is lost. For instance, a non-monotonicity of the transfer characteristic (see Fig. 1.7(b)) cannot be detected: the ADC produces a code word that is not checked with respect to the sampled analog level because no correct digital representation of the latter exists for comparison. For this reason, qualitative testing (see above) should always precede quantitative measurements to make large errors obvious.

Nevertheless, encoding errors can be estimated using the calculated deviations of thresholds and assuming certain analog levels to be digitized (for example, those which are always situated in the centre of an ideal quantization interval). With regard to the slew rate of the input signal, the sampling time-uncertainty can also be determined from these encoding errors.[21]

6.4 Summary

There is no standard technique for testing converters and there are different definitions of the specifications used in data sheets. Often the performance of the devices is characterized mainly by parameters measured under static operating conditions only, thus delivering data that look somewhat optimistic for real applications. The test methods described in this chapter are therefore proposals for the user and the designer of D/A and A/D converters for measuring the device chosen.

All qualitative techniques, which in principle are restricted to observing a signal on the display of an oscilloscope, will also give first-order information on the performance of a converter for dynamic operation, and they will indicate serious errors.

Quantitative measurements must follow. There is a variety of methods for static testing of ADCs and DACs. The set-ups usually employ a mini- or microcomputer for handling the digital data and controlling the test run.

Although DACs deliver an analog output signal and well-known analog techniques of measurement can be used, it is difficult to measure dynamic performance. For example, a 4-ns full-scale settling time of a 10-bit converter means a settling to within 0.5% of the final value. Commonly, the differential non-linearity (static), the updating rate (which is usually limited by settling time, not by set-up times of digital circuits inside the DAC), and glitches are specified. To the knowledge of the authors, the signal-to-noise ratio of a sine wave at the output of a DAC, fed with an appropriate digital input signal, is not yet indicated in the data sheets.

With increasing availability and application of high-speed ADCs, there is a need for dynamic testing. Back-to-back measurements, using a DAC for reconstruction of the digitized test signal, are only useful for special applications. For more objective tests, the performance of an ADC must be derived from its digital output. Due to high sampling rates, a computer cannot be connected directly to a device under test. A buffer memory must be inserted between the device and the computer or a strobing technique used to reduce the word rate to the input of the computer.

State-of-the-art A/D converters exhibit a sharp decrease in performance for increasing input frequency. For example, high-speed ADCs of the all-parallel type yielding a resolution of 8 bits at maximum word rates of 30 to 100 or 200 MHz may show a loss of signal-to-noise ratio (SNR) from 3 dB (static) to 24 dB at Nyquist frequency, thus maintaining 7.5 to 4 effective bits. When SNR or the number of effective bits are calculated, single large errors are subjected to averaging. Therefore the usefulness of these parameters is restricted to applications such as communications. More stringent performance criteria are differential non-linearity and deviation of thresholds or integral linearity (see section 6.1). Peak values in the order of 300% or four times one quantum Q or more are associated with the decrease in performance cited above.

These numbers given refer to a full-scale loading at Nyquist frequency for test. They may be considered as worst-case specifications due to the characteristics of real world signals which exhibit small amplitudes at high frequencies (Fig. 2.4).

Dynamic performance changes primarily with the slew rate of the input signal. Depending on the structure of the A/D converter under test, this may be interpreted to be caused by aperture uncertainties of an internal sample-and-hold, of the comparators (in a flash-type converter), or by inadequate delays in a pipelining structure, etc. Second, increasing sampling rates can affect performance due to internal pulse feedthrough and to the growing difficulties in latching valid data inside the converter.

In all the tests described, the converters are considered as black boxes. Their errors are detected or measured, but not the sources of errors. These can be referred to the structure and the circuits of the device under test if those details are known. Slight modifications of the testing procedures and additional measurements will help to determine the error sources and their influence on the overall performance of the converter, and a redesign will be possible. For example, if a sample-and-hold circuit provided in an ADC is accessible, its performance characteristics (section 5.2) can be measured separately.[33,34]

Encoding errors of ADCs could be found directly if the testing signal is generated via a DAC whose input is compared with the output of the digitizer under test. This pure digital technique is not treated here because of the problems associated with the quantized shape of the test signal. Only static and pseudo-dynamic measurements can be made. The analog signal exhibits no slew rate in the sampling point, and the sampling frequency is limited due to the finite settling time of the DAC output.[21]

None of the methods of measurement described will deliver all the parameters that are needed for complete characterization of ADCs. One has to decide which test will best fit the requirements of the application. One criterion may be the instrumentation already available in a laboratory. Tests such as sine-wave fitting and a histogram test which can determine the errors of the transfer characteristics promise to give more insight into a converter than others such as DFT techniques. The histogram test even allows a conclusion for the conversion principle (if it is unknown) from the histograms measured which exhibit structures that are characteristic of the digitizing procedure.

Due to the growing use of A/D and D/A converters in systems and instrumentation there is an increasing need for adequate testing of converters. In view of the time and precision required, computer-aided methods promise to be the best approach.

References

1. Pratt, W. J., 'Don't Lean on A/D Spec's ...', *Electronic Design* **22**, 80–84, April 12 (1974).
2. Jacobs, R. W., 'Specifying R. W., 'Specifying and Testing Digital to Analog Converters', *Applications Bulletin AN-25*, Teledyne Philbrick, July (1976).
3. Sheingold, D. H. (ed.), *Analog-digital conversion notes*, Analog Devices Inc., Norwood, Mass. (1977), Chapter 11-4.
4. Tewksbury, S. K., *et al.*, 'Terminology Related to the Performance of S/H, A/D, and D/A Circuits', *IEEE Transactions* **CAS-25**, 419–26, July (1978).
5. Naylor, J. R., 'Testing Digital/Analog and Analog/Digital converters', *IEEE Transactions* **CAS-25**, 526–38, July (1978).
6. Souders, T. M., and Flach, D. R., 'An Automated Test Set for High Resolution Analog-to-Digital and Digital-to-Analog Converters', *IEEE Transactions* **IM-28**, 239–44, December (1979).
7. Pastoriza, J. J., *A Dissertation on Specifying, Measuring and Using very High-speed Digital-to-Analog Converters*, Analog Devices Inc., Norwood, Mass. (000).
8. Duttweiler, D. L., and Messerschmitt, D. G., 'Analysis of Digitally Generated Sinusoids with Application to A/D and D/A Converter Testing', *IEEE Transactions COM-26*, May (1978).
9. Gordon, B. M., 'Noise-effects on Analog to Digital Conversion Accuracy', *The Analogic Data-Conversion Systems Digest* 187–207, Analogic Corp., Wakefield, Mass. (1978).
10. Jacobs, R. W., 'Specifying and Testing Analog to Digital Converters', *Applications Bulletin AN-24*, Teledyne Philbrick, July (1976).
11. Connolly, J. J., Rosenbaum, M., and Rittenhouse, L., 'Critical Design of Parameters and Test Method of Ultra High Speed Analog to Digital Converters' *Digest of the Journées d'electronique 73*, Conferences on Conversion A/D and D/A, Lausanne, 16–18 October, 1973, 373–411.

12. Larson, L., 'Test Subtle ADC Characteristics Using only a Scope and DVM', *EDN* **26**, 97–100, October 28 (1981).
13. Pretzl, G., 'Messung der Fehlerraten von Analog/Digital-Umsetzern', *Nachrichten Elektronik* **36**, 24–9, January (1982).
14. Burney, W., 'High Resolution Converter Cuts Linearity Test to 12 Seconds', *Electronics* **54**, 142–6, September 22 (1981).
15. McClellan, P., *Static Performance Testing of Digitizers*, Program ADSPAN, Tektronix Signal Analysis Group, 30 November, 1977.
16. Havener, R., 'Catch Missing Codes...', *Electronic Design* **23**, 58–64, August 2 (1975).
17. Corcoran, J. J., Hornak, Th., and Skov, P. B., 'A High-resolution Error Plotter for Analog-to-Digital Converters', *IEEE Transactions* **IM-24**, 370–74, December (1975).
18. Pretzl, G., 'A New Approach for Testing Ultra Fast A/D-Converters', *ESSCIRC '77 Digest of Technical Papers* 142–5, VDE-Verlag, Berlin (1977).
19. Koller, H. U., 'New Criterion for Testing Analog-to-Digital Converters for Statistical Evaluation', *IEEE Transactions* **IM-22**, 214–17, September (1973).
20. Pretzl, G., 'Dynamic Testing of High-speed A/D-converters', *IEEE Journal SC-13*, 368–71, June (1978).
21. Pretzl, G., 'Dynamische Eigenschaften schneller Analog-Digital-Umsetzer', PhD thesis, Universität Erlangen-Nürnberg (1978).
22. Smith, B. F., 'High-speed A/D–D/A Converters: Their Effect on Video Signal Fidelity', *Digest of the Journées d'electronique*, Conferences on Conversion A/D and D/A, Lausanne 16–18 October, 1973, 413–38.
23. Kerster, W. A., 'Characterizing and Testing A/D and D/A Converters for Color Video Applications', *IEEE Transactions* **CAS-25**, 539–50, July (1978).
24. 'Dynamic Performance Testing of A to D Converters', *Hewlett-Packard Product Note 5180A-2*.
25. Chin-Moh Tsai, 'A Digital Technique for Testing A–D and D–A Converters', Master thesis, University of Utah, June (1953).
26. Polge, R. J., Bhagavan, B. K., and Callas, L., 'Evaluating Analog-to-Digital Converters', *Simulation* **1975**, 81–6, March (1975).
27. Pratt, B., 'Test A/D Converters Digitally', *Electronic Design* **23**, 86–8, December 6 (1975).
28. Bennett, W. R., 'Spectra of Quantized Signals', *Bell System Techn. Journal* **27**, 446–72, July (1948).
29. Ochs, L., 'Measurement and Enhancement of Waveform Digitizer Performance', *1976 Electro Conference Record*.
30. Ochs, L., and McClellan, P., *Dynamic Performance Testing of Digitizers*, Program MDPMPN, MDPANL, Tektronix Signal Analysis Group, 10 November 1977.
31. Gardner, K., and Johnson, P. T., 'Automatic Aperture Performance Measurement for High-speed Analogue–Digital Converters', *Electronics Letters* **16**, 741–2, September (1980).
32. Cattermole, K. W., *Principles of Pulse Code Modulation*, Iliffe Books, London (1973), Chapter 3.5.
33. Michaels, S. R., 'Mend Flash-converter Flaws with a Track/hold Cure', *EDN* **26**, 109–13, September 30 (1981).
34. Johnston, R., 'Analyzing the Dynamic Accuracy...', *Electronic Design* **21**, 80–83, September 13 (1973).
35. Mahoney, M., 'Automated Measurement of 12 to 16 bit Converters', *1981 IEEE Test Conference, Paper 11.5*, 319–27.
36. Kawayachi, N., et al., 'An Ultra Low Glitch, 50 MHz High Speed 10 bit Digital-to-Analog Converter', *ESSCIRC '82 Digest of Technical Papers*, 65–8.
37. Souders, T. M., and Flach, D. R., 'An NBS Calibration Service for A/D and D/A Converters', *1981 IEEE Test Conference, Paper 11.2*, 290–303.

Appendix: Glossary of Terms

Accuracy: The input to output error of a data converter, expressed as a percentage of full scale or in units of the least significant bit. If referred to a standard voltage source, it is absolute accuracy; otherwise it is relative accuracy (referred to converter reference).

Acquisition time: In a sample-and-hold circuit the time required for the hold capacitor to change from its previous value to a new value when the circuit is switched from the hold to the sample mode. Commonly, it is specified for a full-scale change, and the settling time of the output buffer to within a certain error band is included.

Alias frequency: A false lower frequency component in a reconstructed analog signal (S/H or DAC output) due to inadequate sampling rate and/or low-pass filtering.

Analog multiplexer: *See* Multiplexer.

Analog switch: An electronic switch that handles multi-level, i.e. analog, signals.

Aperture delay: Elapsed time in a sample-and-hold circuit from the hold command to the actual end of sampling.

Aperture jitter: Aperture uncertainty (see below) in the special case when the sampling point is fixed with respect to the signal.

Aperture time: The whole 'time window' for a sample-to-hold transition; includes delay and uncertainty.

Aperture uncertainty: Non-constant portion of the aperture time; depends on signal slope and frequency and the signal level at the sampling point.

Bandwidth: For small signals the frequency at which the gain of a circuit or the amplitude of a signal itself goes down 3 dB from its d.c. value. Full-power bandwidth of an amplifier is specified as the signal frequency that can be amplified for a given voltage excursion without distortion due to slew rate limitations.

Bipolar mode: Characterizing an ADC whose input range extends from negative to positive signal levels. Commonly, the input range is offset by one half the full-scale range.

Bit rate: Number of bits times the sampling rate; a communications term.

Companding: Combination of compressing (in an ADC) and expanding (in an DAC) the analog signal range to yield large dynamic ranges.

Compliance voltage: That voltage allowable at the output of a current output DAC without degrading specified accuracy.

Conversion rate: The number of repetitive AD or DA conversions per unit time (word rate).

Conversion time: Time required for an ADC in the worst case to perform a complete conversion.

Data acquisition system: A system processing one or more analog signals by means of analog multiplexers, sample-and-holds, A/D converters to convert them into digital form (for use by a computer).

Data recovery filter: A low-pass filter used to reconstruct an analog signal from a train of multi-level samples.

Decoder: A synonym for D/A converters; used in communications.

Deglitcher: A sample-and-hold circuit sampling the output of a DAC when it has settled and spikes (glitches) have vanished.

Dielectric absorption: The charge on the plates of a capacitor causing the insulating dielectric to change, so that capacitors exhibit a voltage memory characteristic.

Differential (non-)linearity: The deviation of any quantum from its nominal value in an ADC or DAC. Often called differential (non-)linearity error, expressed as a percentage or as a fraction of one nominal quantum.

Digitizer: Synonym for A/D converter.

Droop: Voltage decay of a hold capacitor during the hold mode.

Droop rate: Droop per unit of time.

Dynamic accuracy: Specifies the total error of a converter operating at its maximum signal and/or sampling frequency.

Dynamic range: The ratio of a converter's full-scale range to the smallest nominal difference the converter can resolve.

Effective number of bits: Number of bits calculated from the ratio of signal power to r.m.s. errors of a real ADC.

Encoder: A communications term for an A/D converter.

Encoding error: Difference between the actual digital output of an ADC and the ideal digital representation of the signal level at the sampling point.

EOC: Digital output of an ADC indicating End of Conversion.

Feedthrough: The amount of an input signal that appears at the output of a sample-and-hold when the switch is open, expressed in dB.

First-order hold: A recovery circuit which uses the present and the previous analog samples to predict the signal slope to the next sample.

Flash-type ADC: ADC sampling the digital outputs of comparators instead of using an analog sampling circuit.

Folding characteristic: V-shape transfer characteristic as used in Gray-code converters.

Frequency-to-voltage converter (F/V): A special kind of D/A converter which converts an input pulse rate into an output analog voltage.

Full-scale range (FSR): Difference between maximum and minimum signal level at the input of an ADC or at the output of a DAC.

Gain error: The difference in the overall slope between the actual and the ideal transfer characteristic of a data converter.

Glitch: Output over/undershoot at bit transitions in a DAC due to different switching instants of the weights associated with the bit positions that change.

Hysteresis error: Variation of the analog threshold levels of an ADC depending on the direction from which the levels are approached.

Integral (non-)linearity: The (peak) deviation of a data converter transfer characteristic from an ideal or best straight line with offset and gain errors zeroed. Commonly expressed in fractions of LSBs or as a percentage of FSR. For A/D converters there is also another definition in use: the peak values of the quantizing error exceeding the bounds of plus/minus one half quantum.

Jitter: Uncertainty in time about an event that should happen at a specified instant.

Least significant bit (LSB): The bit with the least weight in a digital word (lowest order bit).

Linearity: *See* Integral (non-)linearity.

Long-term stability: The change of the accuracy of a converter or circuit with time.

Major carry transition: The change of the most significant bit opposite to a change of all other bits at the same time at the input or output, respectively, of a data converter.

Missing code: A code word that never occurs at the output of an ADC when the input signal is varied over the full range.

Monotonicity: Describes a transfer characteristic with the output increasing for an increasing input.

Most significant bit (MSB): The bit with the highest weight in a digital word.

Multiplexer (MUX): An addressable array of switches connecting one of a number of inputs with the output. Depending on the signals to be handled, analog and digital multiplexers are distinguished.

Multiplying DAC: A DAC whose reference voltage can be varied in a wide range.

Negative true logic: A logic system with the more negative of two voltage levels defined to be a logical '1'.

Non-linearity: *See* Linearity.

Non-monotonicity: The opposite of monotonicity.

Nyquist frequency: The rate at which an analog signal must be sampled due to the sampling theorem, i.e. minimum twice the maximum signal frequency.

Offset error: Deviation of a reference point of a transfer characteristic from its nominal value. The reference point may be at zero voltage in a converter with bipolar or unipolar signal range (zero error) or at the most negative level for a bipolar converter.

Pedestal offset: In a sample-and-hold circuit, the amount of offset voltage in the output caused by the hold mode command.

Positive true logic: A logic system with the more positive of two voltage levels defined to be a logical '1'.

Quantizing error: The inherent difference between an analog voltage and its representation after digitizing in an ideal ADC and decoding in an ideal DAC.

Quantum: The analog value of the increment between two adjacent codes from an ADC or DAC (the analog value representing the LSB).

Ratiometric A/D converter: An ADC with a variable reference, i.e. a variable full-scale range, measuring the ratio of the input signal to the reference.

Reconstruction filter: A low-pass filter needed for reconstruction of a sampled analog signal.

Recovery filter: *See* Reconstruction filter.

Recovery time: If an overload occurs at the input of an ADC it needs a certain time to recover and to operate with the accuracy specified.

Resolution: The smallest quantity that can be distinguished by an ADC or generated by a DAC. It is commonly expressed by the number of bits, the smallest signal change, or the smallest change relative to full-scale.

Response: The answer of a device to a specified event.

Sample-to-hold step: The same as pedestal offset.

Sample-to-hold transient: A voltage spike at the output of a S/H caused by feed-through of the digital sample-to-hold command. Transient settles out, no permanent error.

SAR: Successive Approximation Register; needed in a special type of ADC.

Settling time: The time delay between a change of input signal and the corresponding change of the output signal which has to settle within a certain error around the final value.

Skipped code: *See* Missing code.

Slew rate: The maximum rate of change of the signal at the output of a circuit.

Track-and-hold: A S/H circuit which follows the input signal during a long sampling time relative to the hold time.

Tracking A/D converter: An ADC continuously following its analog input and continuously updating its output.

Throughput rate: Same as Conversion rate or Word rate.

Unipolar mode: In this operating mode the analog range of a converter has one polarity only.

Word rate: *See* Conversion rate.

Zero error: *See* Offset error.

Index

absolute accuracy, 178
acoustooptic deflection, 90
acquisition time, 141
actual threshold levels, 193
adaptive or companded delta modulation, 85
adaptive step height, 85
A/D conversion by software, 92
A-law, 72
aliasing, 9
amplitude clipping, 154
amplitude-modulated train, 83
analog bandwidth of an ADC, 199
analog logarithmic conversion, 76
analog shift register, 144
analog switches, 125, 137
antialiasing, 6
aperture error, 150
aperture jitter, 145, 236
aperture performance measurement, 205
aperture time, 142, 146
aperture uncertainty, 142
asynchronous ripple counters, 171
audio attenuator, 117
auto-zero intervals, 67
auxiliary-charge-phase switch circuit, 142
averaging during sampling, 141

back-to-back measurements, 214
balancing steps, 66
bandgap reference, 163
bandwidth, 199
BCD-code-DAC, 112
beat frequency test, 200
best straight line, 179
binary-coded decimal (BCD), 12
bit-at-a-time, 69
bit-by-bit conversion, 48
bit-scan mode, 180
bit slice processor, 97, 206
bootstrap integrator, 166
bucket brigade, 144
buried zener diode, 112, 162

C-2C structure, 112
calculation of $\Delta S/N$, 209
capacitive feedthrough, 148
capacitive voltage divider, 75
carry transitions, 181
cascaded converters, 34, 69
cathode ray encoding tube, 26
CCD shift-register, 112
channel resistance, 134, 136
charge balancing converter, 66
charge coupled devices (CCDs), 90
charge redistribution converters, 67, 121
chords, 73, 117
CMOS switches, 137, 138
codec, 75
codes, 11
coding tube, 26
common emitter transistor circuit, 127
common-mode range, 152
companding, 18, 71
companding advantage, 71
comparators, 149
compensating capacitor, 148
complementary binary, BCD, 12
complementary transistors, 131
compressor, 70
constant current charging, 167
contour effects, 189
conversion rate, speed, 5, 63
conversion time, 5, 197
correction DAC, 119
correction PROM, 119
counters, 171
counting method, 53, 61
count-mode, 180
cumulative frequency, 209
current comparator, 160
current comparison, 33
current mirror, 153
current switch, 131, 133
current switching type D ↑ AC, 114
cyclic converters, 50, 69

221

data acquisition building blocks, 96
data acquisition system, 83
decision noise, 213
decoded D/A converter, 114
decorrelation means, 18
delta modulation, 84
depletion type, 134
deterioration of S/N, 207
deviation of thresholds, 207
differential amplifiers, 46, 132
differential encoding, 22
differential (non-)linearity, 5, 178, 187
digital code converters, 74
digital dither, 185
digital error correction, 39
digital sample and hold circuit, 149
digital switches, 125
diode-bridge, 138
diode bridge sample and hold circuit, 144
direct method, 24
discrete Fourier transform, 202
displacements of thresholds, 188, 196, 207
distortion, 9
dithering, 89
dither pattern test, 190
DPCM encoding, 84
DPDT, 137
dual ramp (slope) converter, 63
dual rank flash*flash, 49
dummy switch, 148
dynamic crossplot, 190
dynamic programming of digital input, 180
dynamic range, 77
dynamic testing of analog-to-digital converters, 197
dynamic testing of digital-to-analog converters, 181

Ebers and Moll, 128
effective bits, 205, 209
efficient coding, 15
electrooptic, 90
electromechanical switches, 124
electron beam semiconductor target (EBS), 90
encoding errors, 205, 213, 215
enhancement type, 134
entropy, 16
error band, 181
error correction in D/A converters, 119
error plot, 193
estimation, 84
exclusive-OR gates, 27

expandor, 70
exponential gain amplifier, 82
extended counting method, 53, 61, 64
extended parallel technique, 54

fast Fourier transform testing, 204
feedback approach for testing, 192
feedback converters, 36, 50
feedback system, 148
feedforward, 36
feedforward amplifier, 48
feedthrough, 142
field effect transistor as a switch, 133
figure of merit, ADC, 54
finite gain, 152
first order low pass response, 140
flash mode, 25
flash type converter, 149
flat-top sampling, 140
floating point A/D converters, 75
folding law, 41
frequency distribution, 195
frequency divider, 172
full power bandwidth of an ADC, 199
full-scale error, 178, 180
full-scale output, 178
full-scale range (FSR), 7
full-scale settling, 182

GaAs field effect transistors, 90, 148
gain-bandwidth product, 152
gain error, 5, 53, 152, 180
gain or full-scale error, 178
glitch, 185
glitch charge, 185
glitch energy, 181
glitches, 114
Gray Code Converters, 41

high-speed comparators, 154
hold capacitor, 140
hold droop, 197
hold mode, 68
hot-carrier diodes, 127
hysteresis, 189
histogram test, 206

ideal switch, 124
idle or granular noise, 85
improvement of the S/N ratio, 71
indirect converters, 24, 61
information theory, 15
input/output characteristic, 176

instantaneous signal-to-noise ratio S/N, 70
integral (non-)linearity, 179, 188
integration during sampling, 141
interleaving of ADCs, 91
interpolation A/D converter, 58
interpolation filter, 108
interpolative coders, 22
interpolative D/A converters, 122
interpolative nonlinear converters, 75
irrelevant messages, 15

jitter, 145
Josephson junction (JJ), 90, 160
junction FETs, 134

ladder-network D/A converters, 110
Laplacian PDF, 10
latching comparator, 154
Least Significant Bit (LSB), 6
level-at-a-time, 53, 61
linearity error of the Miller Integrator, 166
loading factor, 10
logarithmic conversion, 76
longtail pair, 132
loss of bits, 209
low pass filter, 6, 9

major carry transition, 113
Markoff-source, 18
maximum throughput rate of ADC, 197
maximum word rate of DAC, 186
Max-quantizer, 18
mean square distortion measure, 10
measurements on D/A converters, 177
mechanical switches, 125
memory mapped input/output, 96
mid tread, 189
midriser, 189
midscale transitions, 185
Miller-integrator, 165
minimum mean square error (MMSE), 18
missing (skipped) code , 5, 193
monolithic feedback converters, 56
monolithic sample and hold circuits, 147
monotonicity, 5, 65, 180
MOS-comparator, 155
MSFET↑s, 134
Most Significant Bit (MSB), 6
multiple folding, 48
multiplexed A/D converters, 83
multiplying D/A converter, 117

natural sampling, 140

N-channel, 134
negative true logic, 12
noise measurements, 193
noise power, 9
noise shaping, 22
noise suppression, 65
noisy transitions, 186, 189
nominal threshold voltages, 193
non-linear A/D converters, 70
non-linear D/A converters, 74, 114
non-linear quantization, 60
non-monotonicity, 114, 197
non-uniform quantization, 70
number of events, 194
Nyquist frequency, 8, 18

offset error, 5, 178
ohmic region, 135
ON-resistance, 136
one bit per cycle, 51
one's complement, 14
optoelectronic switches, 133
opto-isolator switch, 133
overload distortion, 10
overload noise, 85
oversampling, 22, 60

parallel converter, 24
parallel switch, 125
pattern generator for testing, 180
P-channel, 134
peak linearity error of DAC, 180
Pinch-off voltage, 134
pipelining converters, 49
position encoding, 3
positive logic, 12
post subtractive conversion technique, 34
predictive coder, 18
predictor, 84
presubtractive conversion technique, 50
priority encoding, 30
probability density function (PDF), 10
probability distribution, 194
probability of the codes, 186
programmable voltage source, 191
propagation converter, 34
pulse code modulation (PCM), 4, 8, 70
pulse feedthrough, 197
put and take technique, 35, 69

quantized feedback converter, 66
quantizing characteristic, 186
quantizing interval, 178, 180

quantizing noise, 9, 203

ramp function, 61
rate distortion function, 18
redistribution mode of an ADC, 68
redundancy, 17
reference DACs, 181
reference signal (estimated), 204
reference spectrum, 203
reflected binary code, 26
relative accuracy, 178
relative error of floating point ADC, 77
R-$2R$-resistive ladder, 110
resolution, 35
reverse operation of transistors, 130
ripple counters, 171
ripple through converters, 34
robust quantization, 70

sample-and-hold circuit, 6, 139
sampling jitter, 213
sampling-on-the-fly, 149
sampling period, 108, 140
sampling pulse with finite width, 140
sampling technique for testing of ADCs, 201
sampling theorem, 8
sampling time-uncertainty, 213
saturated emitter follower, 131
saturation region of transistor, 127, 131
sawtooth generators, 164
sawtooth shape of quantization error, 188
Schmitt trigger, 158
selfcalibrating 16-bit D/A converter, 192
semiconductor diode, characteristic of, 126
sequential-parallel A/D converters, 34
series/parallel/series structure, 91
series switch, 125
servo technique for A/D conversion, 51
Shannon–Rack decoder, 120
shunt diodes for overload protection, 157
sigma–delta-modulator, 66
sign bit, 13
sign-magnitude codes, 14
signal power, 9
signal-to-noise ratio (S/N), 9, 196, 203
simultaneous method, 24
sine wave fitting test, 204
single-ramp method, 61
slope of quantizing characteristic, 187
software, A/D conversion by, 92
SPST, 137
staircase function at the output of a DAC, 108

staircase generators, 170
static testing of digital-to-analog converters, 180
stochastic A/D converters, 86
stochastic D/A converter, 60, 122
straight binary code, 12
strip-chart record, 189
stripline sampling technique, 91
strobe facility of comparators, 154
strobing technique for ADC testing, 207
subranging converters, 34
successive approximation converter, 36
successive approximation register (SAR), 51
superposition errors within DACs, 180
symmetrical stage-design of Gray Code converter, 44
synchronous counter, 173

temperature compensation of a Zener diode, 162
terminating resistors of a ladder network, 111
test pattern for DACs, 181
testing converter, 177
threshold voltage of quantizing interval, 151, 187
thyristor, sawtooth generator with, 165
time interleaving of ADCs, 50
time sharing of an ADC, 83
time uncertainty in a S/N, 146
T1-system, 70
track-and-hold, 142
tracking converter (ADC), 57, 197
tracking frequency, 197
transconductance amplifier in Gray Code stage, 44
transfer characteristic, 5, 178, 195
transferred electron devices (TELD), 90
transient spikes of DACs, 183
transient characteristics of comparators, 152
transition points between quantizing intervals, 193
trapezoidal test signal, 207, 212
triple-ramp converter, 64
tunnel-diode threshold detector, 37, 159
two-phase switch, 142
two's complement, 13

ultra-high-speed converters, 89
unit-increment code, 26
updating rate of DAC, 214
up/down counters, 173

μ-law, 71

VAROM, 29
vertical A/D conversion, 97
virtual ground, 157
voltage comparator, 153
voltage comparison, ADC with, 28
voltage divider D/A converter, 114
voltage droop, 142
voltage references, 162
voltage selector, 48
voltage-to-frequency (V/f) converters, 65
volt-equivalent of temperature, 46

V-shape (folding law), 41

weighted current D/A converter, 108
weighted resistor D/A converter, 108
weighting method for A/D conversion, 54, 69
weighting process, 35
word-at-a-time, 24

Zener diode reference, 162
zero calibration for DAC testing, 180
zero-order hold/step interpolator, 108

M. Cunningham 1975.

I hope you will find this book interesting. Kenneth

THE
BIRD GARDENER'S BOOK

THE
BIRD
GARDENER'S
BOOK

Rupert Barrington

WOLFE PUBLISHING LIMITED
10 EARLHAM STREET LONDON WC2

© Rupert Barrington 1971
3rd Impression 1974
ISBN 72340436 4

PRINTED IN GREAT BRITAIN
BY EBENEZER BAYLIS AND SON LTD
THE TRINITY PRESS, WORCESTER, AND LONDON

CONTENTS

Introduction   *page* 7

1. Basic Requirements for Survival   13
2. Natural Nesting Sites   19
3. Artificial Nesting Sites   53
4. Nesting Preferences of Various Birds   79
5. Food, Water, Perches and Roosting   91
6. The Enemies of Birds   105

Conclusion: Bird Conservation   123

Cover photograph: Marsh Tit on a beech bough, by Dennis Avon

INTRODUCTION

THERE ARE thousands of gardens today which could be converted into miniature sanctuaries for birds if the right conditions were provided.

This book attempts to explain how this is done, with particular emphasis on the selection and treatment of the trees and shrubs which are most useful to garden birds. Many gardens have been planted with trees which are quite unsuitable for nesting birds, but there are also many gardens which are in the process of being planted, and it is hoped that some of the suggestions in this book will be especially helpful to those who are planning a garden and who are interested in birds.

The book is based upon observation over a period of five years on the nesting, feeding and roosting activities of birds in a large garden on the edge of a town where, owing to the large number of predators, young birds required constant protection, and where an increase both in the number and variety of garden birds was only achieved after several seasons.

Why Preserve Garden Birds?

It may well be said that there is no shortage of these birds. This is true of rural areas, but it is not true where large, treeless housing and factory estates have been built. Here there are only sparrows and starlings—and as the process goes on and the countryside becomes more and more urbanized, so will the more interesting species of garden birds vanish.

Slow and creeping changes in our environment are occurring

Introduction

all the time. These changes are either deliberate or accidental.

A good example of a deliberate change is the situation which arises when a building contractor buys up an old Victorian house with a large garden containing many trees and shrubs in which birds have nested regularly for years. The site is bulldozed clear and perhaps thirty new houses are built on it. Each owner of a new house may have about a quarter of an acre of land where he will no doubt make a small lawn and plant a few flowers with possibly a low dividing privet hedge. His neighbour may do the same. In this way what used to be a favourite breeding ground for songbirds such as the robin, thrush, blackbird, mistle thrush, goldfinch, linnet, wren, and many others, has been converted into an area in which only sparrows, starlings and a limited number of other birds which are good colonists, such as the blackbird, will breed. Most of the original bird inhabitants will have to move elsewhere to find breeding grounds.

An example of an accidental change in environment is the planting of many thousands of acres of non-agricultural land in England, Scotland and Wales with conifers for commercial use on ground where deciduous trees and undergrowth grew before. Although this may at first seem a wonderful thing for birds, in fact it is not because conifers generally have a much poorer associated wild life than the broad-leafed trees—and this applies expecially to the types of conifers which have been imported from abroad. Another reason is that all wood to be used commercially has to be straight, and to get trees to grow straight they must be planted very close together. Thus, when the young conifers have reached a certain stage in growth their branches are so close together that very little light can reach the ground beneath, with the result that there is no plant growth or insect life below. No ground feeding birds can live under such conditions, and the only birds which could benefit would be the relatively few which feed on the seeds or shoots of conifers. Although the latest practice is to space conifers more widely to reduce thinning costs, it is doubtful if this would allow more than a rather sparse undergrowth to grow once the trees had reached a certain height.

Naturally-occurring forests on the other hand—such as the New Forest—have a very rich undergrowth and many open spaces

Introduction

with gorse, bracken and bramble which have grown without man's interference for hundreds of years. Such surroundings are ideal for certain species of birds, provided their privacy during the breeding season is not disturbed by the weekend invasions of picnickers—something which is now happening all too often.

Dramatic and sudden changes in environment can be caused by man. A very good example of this was the introduction of myxomatosis, which was intended to affect one species only but which in fact had a devastating and far-reaching effect on wildlife in general. It is still quite impossible to assess the damage it caused to species other than the rabbit. This all happened within a very short time and a similar thing could happen again.

This may not seem very relevant to garden birds, but it serves to illustrate how much commercial interests have interfered with the environment of birds, and to show that nearly all types of wildlife are now under a continuous pressure. Their natural habitat is being taken away and their lives endangered.

We must try to remedy this by giving back some of the things we have taken away. This applies to all gardens, big and small, but more particularly to the newly-established garden on an estate where birds have been ousted from their previous nesting places. Birds need tree and plant life to survive, and if instead of laying down concrete, labour-saving slabs we try to provide them with their natural surroundings, we are likely to be rewarded by their presence in the garden.

Gardening usually begins as a necessary chore, but often finishes as an interesting hobby, and if I had not been forced to spend my weekly half-days gardening, I should not have had the material for this book.

Having always had a great interest in birds, I usually tried to direct my gardening activities so as not to interfere with their nesting activities; but this was in a rather negative way, such as not cutting hedges till all nesting activities had stopped. It was not until I allowed a hawthorn road hedge one metre high to grow in stages up to a height of two metres (to prevent people looking over) that I realized how easy it was to create nesting sites for garden birds; for in this 60 metres of hedge about six species of birds began to nest each year.

Introduction

In the many housing estates which I visited, I had noticed the scarcity of garden birds and the absence of suitable trees, and I decided to try out on a larger scale the idea of providing nesting sites, both natural and artificial. I was lucky enough to find a house in Cheshire with a large garden ideal for the purpose, because it was full of evergreens and broad-leafed trees which had not been looked after for many years, with the result that the broad-leafed trees had outgrown the evergreens, depriving them of light and stunting their growth. This presented a very good opportunity to demonstrate how medium-sized trees could be made into natural nesting sites by cutting or pruning, and how, by felling or pollarding larger trees, they could be made more useful to birds.

In the first three years everything seemed to go wrong. The existing bird population was small. There were, for example, no finches except a few greenfinches, no mistle thrushes except as winter visitors, no flycatchers, few blackbirds, and a very few thrushes. Robins, wrens and hedge sparrows were rarely seen, but predators in the form of cats, grey squirrels and magpies were everywhere.

In the fourth year, after much hard work on the trees and shrubberies, things began to take a turn for the better, and at the same time the garden had been made more predator-proof. In that year, some lesser redpolls and goldcrests appeared and nested successfully.

After a bad start, when most of the April nests were destroyed, the fifth year was most successful (the criterion of success being that the young birds were seen flying strongly with the parent birds) and species new to the garden, including chiffchaffs, blackcaps, stock doves, goldfinches, and mistle thrushes, all bred successfully. The more common garden birds, with the exception of hedge sparrows and chaffinches, all had excellent breeding seasons, judging by the number of young birds seen in August. Spotted flycatchers, for example, reared eight successful broods. Tree creepers, nuthatches, long-tailed tits and great spotted woodpeckers have not nested so far, but they are more regular visitors.

It is hoped that next season will be equally successful. In the

Introduction

meantime, it will be interesting to see how many extra birds there are to feed during the coming winter. On the whole the scheme has worked very well, protection from predators being the key factor in this area.

CHAPTER ONE

Basic Requirements for Survival

BIRDS NEED to be extremely adaptable creatures to survive a changing environment and the constant threat to their existence from enemies. Some are more adaptable than others, but even these can survive and reproduce their offspring only if the following four basic requirements are fulfilled:

(1) Suitable nesting sites (see Chapters Two, Three and Four).
(2) A regular food and water supply (see Chapter Five).
(3) Suitable roosting places (see Chapter Five).
(4) Freedom from persecution (see Chapter Six).

Each condition is as important as the other. We may put out large quantities of food for birds during the winter, but if there is nowhere for them to nest during the spring they will either go where conditions are suitable or will be unable to rear any young birds, with the result that the local bird population decreases. Similarly, if they have ideal nesting conditions but are persecuted during the breeding season, their numbers will also decrease.

These are the four conditions vital to a bird's existence and to the reproduction of its kind. To fulfil all these requirements is difficult but by no means impossible. If we can do so, birds will certainly arrive as if out of the blue and will become garden residents. There is usually no difficulty about the food supply, but the other three conditions are more difficult to achieve and it will take a little time to get the desired results. However, it will be well worth the trouble and the presence of birds in the garden will add greatly to its attraction.

Anyone who has a small suburban garden may think that birds will not come to a heavily built-up area. The answer to this is that birds of every species, except when actually nesting, are always on the look-out for suitable localities in which to live and breed. Many of our summer visitors, for example, on their way back from Africa, fly in thousands straight over London and its vast suburbs on their way to find breeding grounds and they drop down here and there for food and water. If they find the conditions suitable, they will settle down and breed and may return again in the following year. During World War II a rare bird called the black redstart settled down and nested among the weeds and rubble of the bombed-out buildings near St. Paul's Cathedral and wild ducks nested near the water tanks used for fire-fighting. This shows that birds will always appear as if from nowhere, if we create the right conditions for them.

Thus we need not worry that birds object to bricks and mortar, but we do need to worry about the rapidity with which we are driving birds from their natural surroundings while not providing them with alternative accommodation suited to their needs. This gradual encroachment on the natural habitat of birds is going to continue—slowly but surely. The use of chemicals on agricultural land has come to stay. The countryside is being invaded more and more by the motor-car, and the air by the aeroplane. Many of our rivers and lakes are no longer peaceful, but resound with the din of the speedboat or outboard motor—and this especially at the height of the breeding season. The pollution of rivers is causing national concern not, it seems, because we are particularly worried about the creatures who live in or near the rivers, but because pollution might have a damaging effect on public health. In short, the outlook for wild birds in this country is rather bleak.

Of course, some people may say why bother, birds are a nuisance in the garden, anyway. This may be true of house sparrows, but otherwise it is rather like saying that rain is a nuisance—it can be, but we could not do without it. In the same way birds do a vast amount of unseen good by consuming beetles, slugs, snails, wireworms, greenfly, caterpillars and a host of other insects which are garden pests.

Knowing that birds will readily adapt themselves to a new environment, we must ourselves try to create the four necessary conditions if we have a garden, for it is in such a place that we can to some extent supervise their breeding activities. As already mentioned, birds will nest close to a house if conditions are suitable, and if the occupants encourage them they will become more tame and never go very far away. Their offspring will tend to stay in the same district or even in the same garden provided the numbers do not increase to such an extent that competition for food results.

The birds with which we are concerned, namely the songbirds and other small birds which normally inhabit a garden, have from very early times always nested in trees, shrubs, thick undergrowth, on steep banks and in holes in trees and cliffs. When man began to build houses, many birds such as the swallow, house martin, sparrow and starling found that human habitations provided just the right sort of nesting environment. There was usually water nearby and waste food was often available. There was no sewage disposal, so that flies were plentiful. Corn growing and other cultivation generally made food supplies more readily available. Most important of all perhaps was that these birds found the nearness of man afforded some protection against their natural enemies in the air, the falcon, hawk and crow, and those on the ground such as the fox, weasel, stoat and wild cat. Those that would not adapt to building their nests actually in man-made constructions, often nested very close to buildings in nearby trees and shrubs.

Sparrows, starlings, swallows, house martins and swifts have benefited by the multiple nesting sites which man has provided and as a result their numbers have greatly increased. Our garden birds, with the exception of the starling and probably the sparrow, were originally woodland birds. House martins and swifts still nest in cliffs if there is no more suitable nesting site available. On the Continent swifts sometimes nest in tree holes. Swallows were probably cave nesters but are also supposed to have nested in hollow trees, flying in through the top, as in an old-fashioned, wide chimney, and they may even have nested on the ends of horizontal boughs.

During the past 250 years, the cultivation of gardens and the Englishman's love of fencing off his property with hedges have been perhaps the main reasons why England remained so rich in bird life for so long. The English hedge is by no means the most economical way of fencing a field, but it is certainly the most picturesque and, from a long-term agricultural point of view, by far the best way of preventing soil erosion and giving protection to livestock. Some farmers on the light land in East Anglia have suffered great loss of soil from dust storms due to the lack of windbreaks in the form of hedges and trees; another example of man having his knuckles rapped for trying to be too clever with nature.

Even so, the old English hedge, so important to bird life, is certainly on the way out. Prairie farming is the new thing and it demands that existing hedges be bulldozed out to make fields bigger and easier to farm. Hedges take time to grow and also take up space, while a barbed wire fence with concrete posts can be erected in a matter of hours and will last for many years without maintenance. It will not harbour weeds or vermin and it is a much cheaper short-term investment and a better barrier.

Birds too have a housing problem now, and if a bird cannot find a suitable nesting site at the right time its breeding efficiency will suffer considerably in that particular season. The reason for this is interesting. The egg production mechanism of most birds is not stimulated to activity unless the hen bird goes through the process of nest building. The ovulatory process seems to be stimulated by the handling of the nesting materials, and when the nest is completed egg production starts almost immediately. In the early part of the season, there is often an interval of a few days before the first egg is laid. This interval gets shorter as the season advances.

Different types of birds have different rituals in their nest building, but these rituals do not start until the building site has been decided. Many birds decide well in advance, say one or two months, where they are going to build their nest. Generally speaking, the more prolific the bird is in its breeding habits, the quicker it can decide on a new nesting site if the first nest is a failure. Again, the bigger the bird the longer it takes to choose a nesting site and the more easily it is put off from breeding if anything goes amiss. The golden eagle, for example, lays only two or three eggs, and if this clutch is destroyed will probably not lay again in that season. Such birds as thrushes, blackbirds and the finch family are fortunately very persistent nest builders and will usually raise at least two broods a year, although they may have to build as many as four nests to achieve this; but here again they must be provided with appropriate nesting sites. For hygienic reasons, the majority of birds will never use the same nest again to rear a second brood, but if the old nest is pulled down the same bird may well return to exactly the same site again in the following year, or even in the same year.

If, therefore, we wish to attract birds to our garden and keep them there to breed we must try to fulfil the four requirements of suitable nesting sites, a regular food and water supply, suitable roosting places and freedom from persecution.

Feeding is perhaps the most arduous task, but it can be most rewarding when an interesting bird appears. Those who provide only winter food for birds should start when the first spell of hard weather sets in, and the feeding should be continued regularly until the weather becomes milder, usually about the end of March.

Pairing up usually takes place in the middle of February and if by feeding birds until the middle of March they can be kept in the vicinity, it is most likely that they will not go elsewhere to nest. Early nesters such as the blackbird, thrush and mistle thrush, start thinking about nesting sites quite early in March and nest building starts about the end of March. In an exceptionally mild winter, pairing takes place earlier, and the thrush family, particularly the mistle thrush, may build in February and have young before the end of March. Blue tits will often stake their claim on a nest box in February.

Nest boxes and other nesting devices should therefore be put up, if possible, before the middle of February, though I have on several occasions put them up as late as the end of May, with successful occupation by tits, flycatchers and blackbirds.

In the breeding season, which is in full swing from mid-April

The Bird Gardener's Book

till the end of July, the most important job is to keep a close watch for predators, having an occasional peep at the nest without frightening the hen bird off.

The autumn is the time to think of how the existing trees and shrubs in the garden can be made more attractive as nesting sites, or to plant young trees which can be both useful to birds and decorative to the garden.

CHAPTER TWO

Natural Nesting Sites

BIRDS MUST necessarily be very particular about how and where they site their nests, and what may appear to be a good nesting site to a human may appear quite the opposite to a bird. Of course trees, shrubs and hedges may be cut or pruned to encourage nesting birds, but the effort is likely to be in vain if the site remains unsuitable. The selection of a site by a bird depends upon the following factors.

1. The Locality

Birds spend a lot of time selecting the locality in which they settle. They must be certain that a permanent food supply will be available, especially in the breeding season. A greenfinch, for example, which feeds mainly on seeding weeds, will not settle down in a locality where there are no weeds. Food supply therefore primarily determines the type of bird which will be found in a locality, but it need not be very close to the nest. Birds such as tits and finches will fly hundreds of yards to forage. Indeed, as a general rule, in order not to attract attention birds do not feed in the immediate vicinity of their nests.

Water is usually available in most places, but garden birds certainly prefer to have their local drinking and bathing place near at hand.

Roosting places are not so important in the breeding season

when the trees are in full leaf and the cock bird roosts near his sitting mate.

Birds which are incubating eggs are very liable to be disturbed at night by cats and they will therefore avoid localities where cats abound. Unaccustomed lights will cause the hen bird to fly off her nest in a panic and she will not return until morning when the eggs will be cold and infertile or the young birds, if hatched, will have died of cold. Persistent noise, such as traffic, does not deter birds from nesting provided they have become accustomed to it, but sudden unaccustomed noises or bangs will upset them.

2. Texture of Trees

The twig and branch texture is highly important. A hedge must not be too thick to allow room for a nest. Birds must be able to move about freely within a hedge or tree and the sitting hen bird must be able to escape quickly if in danger. Very closely-clipped hedges do not allow birds either free entry or exit and leave no room for manœuvre inside. Such hedges are almost useless for birds.

Perhaps the most important feature of the texture which attracts birds is the amount of protection it affords against predators. Birds will choose a thorny tree as opposed to a non-thorny tree, to obtain protection against cats and foxes and other nest robbers. It is probable that by the process of evolution the successful breeders were those who chose thorny trees in which to nest, and so the offspring can now recognize the prickly trees by instinct rather than by the painful process of testing trees for their 'thorn value'.

3. Foliage

The foliage of the tree must give sufficient protection from rain and hot sun and good concealment from predators both above and below.

4. Nest Foundation

Tree-nesting birds need first of all a firm foundation which can take the weight of the nest. The foundation must have forked

Natural Nesting Sites

branches at the correct angles to support the sides of the nest. The supporting fork must not be subject to too much movement in the wind, which would loosen the nest and cause disaster. Each supporting branch in the fork must be stiff and at least a year old in growth.

5. Height of Nests

All birds have their own particular choice about the height of the nest. Tree-nesting garden birds, with a few exceptions such as the hedge sparrow and some of the warblers who build quite low down or on the ground, usually choose a height of 1 to 2 metres from the ground. Another exception is the mistle thrush, which often builds at a height of 4 to 6 metres in the exposed fork of a tree in March. This bird is perhaps the biggest of our garden songbirds and can usually drive off crows and magpies. Being an early nester it has to rely on camouflage rather than leaf cover to conceal its nest, and upon height to protect the nest from predators such as cats who do not normally like climbing much above 2 metres high—a height from which they can safely leap to the ground. The goldfinch is yet another exception. It will sometimes build its beautifully-woven little nest at the extreme tip of a branch of a 12-metres-high leafy tree where it is almost impossible to see the nest until the leaves have fallen.

The garden birds which do not use the fork structure of trees for nesting, such as robins, wrens, wagtails, tits, nuthatches and flycatchers, usually nest at a height varying between 1 to 3 metres from the ground. The robin often nests on the ground in a high, ivy-covered bank.

PROVISION OF NATURAL NESTING SITES

To have birds nesting in our gardens, it is very important for us to ensure that trees are not too high and not too low and that they offer the right sort of cover at the correct height. The closely-clipped privet hedge less than a metre high which surrounds so many small gardens is absolutely useless for birds, being too thick, of the wrong twig texture and far too low. In many cases, however, an existing garden tree or shrub may be pruned to remedy some defect and so provide a natural nesting site.

Generally, most trees and shrubs are not nearly as useful to birds if allowed to grow unpruned as when they are trimmed and kept bushy. Where possible, trees must be kept at a medium height—say, 3 metres—so that they can be kept under control by occasional pruning. Over-trimming may produce too thick a branch texture and under-trimming may have the opposite effect. It depends on the type of tree and the rate of growth of the branches. A laurel tree, for instance, is a strong, fast grower, and if left to grow freely offers nothing much for nesting birds. If clipped lightly once or even twice a year, it can be made into a pleasant-looking, tidy tree with plenty of nesting possibilities. Beeches, though useless as large trees, can be made into excellent nesting sites when grown as hedges.

It is necessary when trimming a hedge or tree to cut the branches in such a way that they sprout into a fork, and it is such a fork which will be chosen by tree-nesting birds. A fork should contain not less than three branches, and preferably more. The angle of the fork is important; a bird likes an angle of about 70° facing vertically upwards, allowing enough room for the nest and at the same time providing a secure fixing. If the angle faces sideways, nesting is still sometimes possible: if it is much narrower or much wider than 70°, then nesting will not be possible.

Such specialized pruning may sound too much like hard work, but if a pair of parrot-beak secateurs is used, suitable fork structures in, say, a hawthorn hedge, can easily be fashioned and these will be useful nesting sites for perhaps ten years. Periodic thinning inside the hedge may be necessary. This sort of pruning only applies to trees which grow branches at a suitable angle for nest building. It is a waste of time trying to make nesting sites in any tree which has a growth angle as shown in **B, C** or **D** in the drawing. It is also useless to attempt to make nesting sites in a tree which does not have a vigorous growth or a strong branch structure.

The drawings on page 24 show how a newly-planted hawthorn hedge should be grown and pruned to offer the maximum number of nesting sites for garden birds. Each tree should be planted about 30 cm. apart.

Natural Nesting Sites

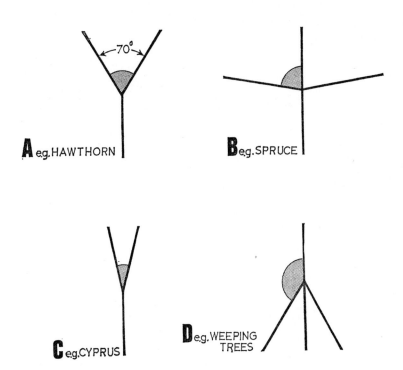

The angle of forked branches is important to tree-nesting birds. In A the 70° angle is ideal. B shows the same angle facing sideways; this might be used for nesting. The angle in C is too narrow, and in D too wide.

The method of pruning as shown should be continued until the hedge is about 2 metres high and 60 cm. thick; when growth is complete the fork structure formed at a height of 1·5 metres will be used as nesting sites by many garden birds.

A nesting fork is obviously of no use to a bird if, as sometimes happens, it has a central shoot growing up the centre of the fork. It is most important to remove this, otherwise no nest can be built.

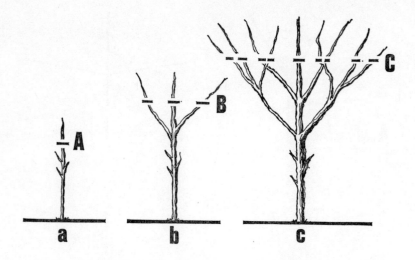

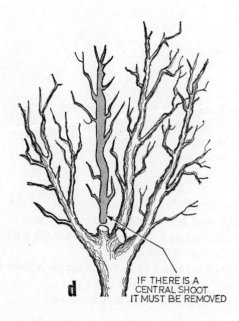

IF THERE IS A CENTRAL SHOOT IT MUST BE REMOVED

Pruning a newly-planted hawthorn hedge. (a) *A two-year-old hawthorn tree has been pruned at* A *in the autumn of its first year of planting.* (b) *Three side-shoots have resulted and these are pruned at* B *allowing about 20 cm. of new growth.* (c) *Further shoots have appeared which are pruned at* C. (d) *A central shoot (shaded) is removed, allowing nesting to take place.*

Natural Nesting Sites

NESTING POTENTIAL OF VARIOUS TREES AND SHRUBS

The trees and shrubs which are commonly found in gardens are listed below alphabetically, and comments are made as to their value as nesting sites. The birds likely to nest in a particular tree are also mentioned. Many trees and shrubs have been omitted either because of their rarity or because they are obviously of no value as nesting sites. The term 'free-growing' which is often used below means that a tree has never been cut back or pruned in any way. This list may seem unduly long and detailed, but our purpose is to find out why some birds seek out a particular type of tree in which to nest and why they never choose what may seem to us to be an excellent nesting site. We might look upon one leafy tree as being as good as another for nesting purposes but perching birds, whose whole life is closely dependent on trees, see the greatest difference between one tree and another, and they are very particular in their choice. To understand why this is so, it is necessary to take each of the common garden trees and shrubs in turn and to study the 'pros' and 'cons' of each, in relation to its value as a nesting site.

APPLE (AND PEAR)

Free-growing apple and pear trees are rarely used by nesting birds. The mistle thrush and the magpie are exceptions to this, the latter being particularly fond of a tall pear tree.

Apple and pear trees when pruned annually and kept at a height of about 3 metres become quite thick and provide good leaf cover. Such trees may well be used for nesting by blackbirds, thrushes, chaffinches, greenfinches and goldfinches. If trained against a wall, blackbirds and thrushes may use them.

Where possible, old apple trees should not be completely cut down, but the branches should be lopped back leaving a stump 2 to 3 metres high. These stumps very often develop holes and cavities in which tits, woodpeckers and tree creepers show a great interest. Artificial nesting holes can be bored in them. They can also be used as a support for honeysuckle, clematis, polygonum, rambling rose or climbing ivy, the latter being used also for roosting.

ASH

This attractive tree is mentioned because it is so common. Unfortunately it has no value as a nesting site for birds. Furthermore it is not a good tree in a garden, for its massive hairy roots spread far out into the surrounding soil and take up all the moisture. Very few shrubs will thrive within 10 metres of it. The best thing to do is to pollard it and cover the trunk with climbing ivy or some other creeper, or to use the trunk for nest boxes.

BEECH

A large free-growing beech is very beautiful at a distance, but in a small garden it can be a nuisance and as a nesting site is of no value to garden birds. In the winter it provides beech mast, which is very popular with chaffinches, great tits, wood pigeons and bramblings, the latter being winter visitors to this country.

Beech grown as a hedge can be most attractive in any garden. It makes a solid boundary fence and does not require much clipping. As a nesting site it is very popular with garden birds, but the hedge should be at least 2 metres high and 60 cm. thick. The thickness of all hedges is very important for nesting birds.

Blackbirds, thrushes, hedge sparrows, greenfinches, chaffinches, bullfinches, and linnets will use the beech hedge as a nesting site.

BERBERIS

The various types of berberis are worth a place in the garden for, even if they are not very good for nesting, their autumn berries are much appreciated by birds.

The *stenophylla* and *darwinii* species are evergreens and very suitable for hedging. If grown up to 1·5 metres high and kept well clipped, they may be used by blackbirds, thrushes, hedge sparrows, linnets and greenfinches for nesting.

BIRCH, SILVER

The graceful silver birch is rarely used by birds except when it becomes old and diseased. Woodpeckers may then excavate a nest-hole in the trunk. Birch leaves often attract greenfly and are therefore a good source of food for tits feeding their young. The

Natural Nesting Sites

seeding heads also attract greenfinches, siskins and redpolls. Birches are subject to a parasitic growth of thick twigs called 'witches' brooms' and these may be used by wood pigeons, thrushes and blackbirds for nesting. The irregular growths and cracks which sometimes occur in the trunk of an old tree attract tits and tree creepers.

BLACKBERRY (CULTIVATED)

A patch of cultivated blackberries at the end of a garden may attract smaller birds such as warblers, linnets, hedge sparrows and yellowhammers. It is essential to allow weeds and grass to grow up inside the patch to give extra cover. In the autumn the growth which has fruited should be cut away, leaving only the new briars.

If grown against a wall, blackberries must be trained against trellis work as they would not by themselves give enough support for a heavy nest such as that of the blackbird.

BLACKBERRY (WILD)

Wild blackberries, as seen on open heath and common land, grow into dense, impenetrable clumps which provide very safe nesting sites for blackbirds, thrushes, hedge sparrows, chaffinches, greenfinches, whitethroats, chiffchaffs, blackcaps, linnets, long-tailed tits, yellowhammers, red-backed shrikes and, in the southern counties, nightingales. Such a formidable list should be enough to persuade any bird enthusiast to set aside a piece of unwanted garden in which to grow the common bramble.

It is necessary to trim blackberry bushes annually to keep them attractive to birds. In fields and open country, this pruning is very adequately done by cattle.

BLACKTHORN

Very few people nowadays would deliberately plant a blackthorn tree in their garden unless they were very fond of sloe gin. In the past blackthorn was much used as a hedging shrub, its murderous thorns making a formidable barrier for fencing in cattle. Its twig structure is in every way suitable for nesting birds and its thorns are a most efficient deterrent to all except the smallest climbing predators. It needs occasional trimming to make good fork structures. Long-tailed tits will often build in a free-growing blackthorn tree. It can be extraordinarily difficult to see the beautiful lichen-covered nest which is built before the branches are covered with leaves, so good is the camouflage of the nest.

Blackbirds, thrushes, chaffinches, yellowhammers, whitethroats, bullfinches, greenfinches and wood pigeons as well as long-tailed tits will use the blackthorn for nesting.

BOX

This tree, once so popular in Victorian gardens, is not seen very often in newly-planted gardens. It was mainly used for ornamental hedging and topiary, and as such the texture is usually too thick for birds.

If box is free-growing it will grow up to 5 metres high as a small tree and will then provide useful nesting sites for the blackbird, thrush, house sparrow, greenfinch, goldfinch, bullfinch, lesser whitethroat and blackcap.

CEANOTHUS

The semi-hardy evergreen type is best for the climate in this country. It should be grown on trellis on a sheltered south wall to a height of 3 metres. Robins, wrens, blackbirds, thrushes, hedge sparrows, house sparrows, flycatchers and pied wagtails may choose it for nesting.

CHERRY (FRUITING AND ORNAMENTAL)

Both types of cherry, for obvious reasons, are free-growing and do not provide sufficient leaf cover for most garden birds except the mistle thrush, which often chooses a somewhat exposed fork in a cherry tree for its nest.

CHESTNUT (HORSE AND SPANISH)

Both these beautiful trees are unfortunately of very little value to garden birds, but like all large-leafed trees they do provide a

Natural Nesting Sites

great deal of shelter from heavy rain and hot sun during the summer. A goldfinch will occasionally nest in the end branches of a horse chestnut tree.

CLEMATIS

All types of clematis can be used to the advantage of birds, mainly due to the leaf cover they provide. They can be trained up trellis work or up the trunk of a dead tree or over an archway. The texture of clematis is not very strong and therefore by itself it would only support the nest of a small bird such as a linnet or goldfinch. With the support of trellis work larger birds will build in it. Some garden hybrids require annual pruning and so are unsuitable.

COTONEASTER

These decorative shrubs, if free-growing, do not have a branch structure of much use to nesting birds, but if grown against a wall or trellis work and kept fairly well clipped to promote thicker growth, they will be used by blackbirds and thrushes.

The berries of most types of cotoneaster are generally much sought after by fruit-eating birds such as the blackbird, thrush, mistle thrush, fieldfare, redwing and waxwing.

CRYPTOMERIA

This very elegant Japanese cedar grows rather slowly to a medium height. It is a good tree to put at the back of a border where a linnet or hedge sparrow may find its bushy growth an attractive nesting site.

CURRANT (FLOWERING)

This shrub, if kept well trimmed each year, is surprisingly useful to blackbirds and thrushes. It is a vigorous grower, seems to thrive on pruning and the twig structure is very suitable for pruning out nesting sites.

ELDERBERRY

The elder is not often seen in gardens as it is looked upon as being too much like a weed. It grows rapidly up to a considerable height. Nevertheless it is quite popular with birds even when free-growing for, though the foliage is not particularly thick, the branch angles are ideal for nest foundation and the bark, being somewhat rough, makes the nest more secure.

To improve the nesting possibility it should be pruned annually, leaving about 30 cm. of new growth each year. In this way good nesting sites can be made especially for blackbirds, thrushes, chaffinches and bullfinches.

The hollow stems of an old tree are popular with tits and tree sparrows and the autumn berries of the elder are eaten by various birds including warblers.

ELM

Large free-growing elms are rarely used by garden birds for nesting purposes. Nevertheless, the trunk of an elm, say, 9 metres high whose top has been cut or blown off after becoming rotten inside, will always be a focus of interest to woodpeckers, tree creepers and wrens in their search for insects which abound when a tree is partly rotten. Woodpeckers may well select such a tree for building. They always use a partly rotten tree, their nesting hole being usually occupied in the following year by a starling, nuthatch, tree sparrow or house sparrow, any one of which will stick to the hole for as long as it can.

The bristly growth of branches from the lower trunk of an old elm may attract blackbirds and mistle thrushes. Jackdaws will invariably nest in the naturally-formed holes at the top of the tree.

Elm bark is so rough that agile climbers such as the stoat can run up and down quite easily. I have seen a stoat removing the young from a jackdaw's nest 10 metres from the ground. It is therefore not a good place to fix tit boxes. Woodpeckers cleverly make their nest hole on an underslope of the trunk where a stoat or squirrel would have a somewhat precarious grip and would be at a great disadvantage if attacked by an angry woodpecker.

Small elm sucker shoots, spreading out from the parent tree into tall rough grass, often contain the nest of a linnet, yellowhammer, blackcap or other warbler. A shoot which spreads from a nearby elm into a garden can be fashioned into a useful nesting shrub.

Natural Nesting Sites

ESCALLONIA

This shrub, which is quick-growing, can be made into an excellent hedge and responds well to clipping—though this, of course, removes most of the very beautiful pink flowers. Blackbirds, thrushes, hedge sparrows, bullfinches, linnets and whitethroats will nest in a hedge of escallonia.

FIG

The fig tree is not often seen except in an old garden against a wall. The twig structure is useless for nesting birds but it can be an ideal place within which to fix nesting structures such as bundles of sticks or an open nesting box, particularly if the wall is high enough to deter predators. Blackbirds, thrushes, robins, hedge sparrows and flycatchers may use such a site for nesting.

FIR

There is such a wide variety of coniferous trees that it is extremely difficult to generalize about their value as nesting sites. It is best perhaps to classify them according to their growth habit, which is of primary importance to the nesting bird.

(1) Those which have a very upright branch growth like many of the cypress trees and have a very dense and compact foliage. These trees are sometimes used by robins and wrens, but bigger garden birds find them too thick for nesting, and the branch angles are usually not suitable for nest foundation.

Old cypress trees with thick trunks are sometimes used by cavity-nesting birds such as stock doves and jackdaws. The cavities occur as the result of the accumulation of many years of dead foliage and twigs between the trunk and branches of the tree. This debris cannot be dislodged by wind or rain owing to the very compact nature of the tree.

(2) Those which have a more or less horizontal branch growth such as the common Christmas tree or spruce. When young and bushy and grown in clumps, such trees are attractive to low-cover nesters such as linnets, hedge sparrows and warblers. When they reach a height of 2 to 3 metres blackbirds and thrushes will use

them. When about 10 years old the lower branches die off and the trees become suitable only for birds which nest at a greater height such as wood pigeons and jays.

Cupressus lawsonii, although it has a rather horizontal type of branch growth, can be made very useful for nesting birds because it responds to clipping very well, producing a thick growth at the ends of the branches.

Cupressus macrocarpa, a very quick growing and bushy conifer is, as a young tree, particularly useful as a nesting site. If the top is cut off when about 2 metres tall, it will bush out and make a good screen and an excellent nesting place for blackbirds, thrushes, greenfinches, hedge sparrows and linnets.

Larches, cedars, deodars, Scots firs and long-needled pine trees have generally a very limited use as nesting sites for garden birds. Golden-crested wrens, however, will sling their tiny nests under the branch of a cedar or deodar tree.

(3) Those whose branches grow downwards from the main trunk. Such trees are of very little value as nesting sites. An example of this group is a Wellingtonia or giant redwood, the trunk of which is a favourite roosting place for tree creepers who find its spongy bark an excellent insulator against cold.

FORSYTHIA

The golden bell variety which has a strong growth but rather meagre foliage will only be useful for nesting birds if grown against a wall or trellis work, where it will attract blackbirds, thrushes, robins, wrens, hedge sparrows, house sparrows and flycatchers.

FUCHSIA

In those parts of the country where the winter is mild, the fuchsia can be grown into a fair-sized bush or made into a very neat hedge by annual clipping. Smaller birds such as linnets, hedge sparrows, yellowhammers and bullfinches will often nest in it.

Natural Nesting Sites

GORSE

There are very few gardens which have a patch of gorse and it is rarely seen except on commons and land not fit for agricultural purposes. This is a pity because it is a most valuable shrub for our smaller birds, and a patch of gorse can be a stronghold which few predators care to penetrate. It is easy enough to grow, once the root is established.

Linnets, whitethroats, golden-crested wrens, long-tailed tits and yellowhammers are particularly fond of gorse for nesting purposes. The rare Dartford warbler invariably nests in a low gorse bush growing in heather.

HAWTHORN

Of all the trees found in the British Isles, hawthorn is by far the most popular with all types of birds for nesting purposes. Fortunately it is still very common, being used almost everywhere for hedges, where its strong, thorny growth makes it ideal. From a bird's point of view it has just the right branch structure if allowed to grow freely, and the leaf coverage is quite adequate. Its thorns are it chief asset and give excellent protection against the larger ground predators.

Hawthorn hedges are often clipped year after year to exactly the same height, with the result that the hedge becomes so thick and woody that it is of no use at all to birds. Such a hedge should be allowed to grow freely for a year and then clipped in the autumn, leaving about 15 cm. of new growth. It must be allowed to reach a height of about 2 metres and it will then be an exceedingly popular nesting site for blackbirds, thrushes, hedge sparrows, linnets, chaffinches, greenfinches, bullfinches, yellowhammers and warblers.

Hawthorn growing freely as a standard tree will look very attractive in the spring with its white blossoms. Its autumn berries are a useful source of food for mistle thrushes, fieldfares, blackbirds and redwings. It is used for nesting by thrushes and blackbirds, chaffinches, goldfinches, wood pigeons and even magpies.

Strange though it may seem, road hedges are very popular nesting sites. This is partly because so many hedges consist mainly of hawthorn, but it may also be due to the fact that birds find there are fewer predators near the roadway. The headlights of cars at night will keep away many of the nesting birds' nocturnal enemies and the traffic during the day will act as a deterrent to daylight robbers.

The birds which breed in road hedges include chaffinches, greenfinches, linnets, yellowhammers (which are now becoming rarer), the warbler family, blackbirds, thrushes, cuckoos, hedge sparrows, red-backed shrikes (now very rare), pheasants, partridges and many others.

HAZEL NUT

A free-growing hazel nut tree is of little value to garden birds, with the exception of the turtle dove. Even if grown as a hedge it does not make a particularly good nesting site except for yellowhammers and bullfinches. In hazel nut plantations the bushy growth around the stocks may sometimes contain the nest of a nightingale or robin. Hazel nuts will, of course, be appreciated by woodpeckers, tits and nuthatches.

HEATHER

If the soil is acid, certain types of heather can be grown up to 1 metre high with great advantage to such birds as the willow wren, whitethroat, chiffchaff and blackcap. Heather is an excellent weed excluder and requires very little attention.

HOLLY

There are quite a number of different kinds of holly; some of these are extremely prickly, while others are not. The prickly type has a very tough twig texture. If grown as a closely-clipped hedge it is almost impossible for a bird to get in and out and equally impossible for it to move around inside, and a hedge of this sort will harbour very few nests. But if free-growing, blackbirds, thrushes, greenfinches, house sparrows, wood pigeons and collared doves will use it for nesting. Blackcaps will nest in small, young bushes.

The non-prickly kinds, if free-growing, tend to have rather drooping branches, but with judicious cutting back they can be

Natural Nesting Sites

made thicker. Mistle thrushes, collared doves, blackbirds and thrushes are likely to use such trees.

Hollies, though slow growing, are well worth a place in the garden, the variegated types being especially attractive. From the birds' point of view they afford protection from the weather all the year round and in winter their berries are a great stand-by.

HONEYSUCKLE

There are several forms of this, and it is better to select one which makes a vigorous growth. In order to keep it bushy it must be vigorously pruned each year, cutting back the long, new shoots by about half their length.

It may be trained on a wall trellis or, better still, over a rustic arch about 2·5 metres in height, after wire netting of 2·5 cm. mesh has been put across the top of the arch. A good place to train honeysuckle is round the stump of an old tree, thus converting what may be an eyesore into something pleasant to look at. Evergreen honeysuckle grown against a house wall makes a good roosting place.

If the honeysuckle is thick and nest support is strong, most garden birds will use it for nesting.

HORNBEAM

These are usually seen as large, free-growing trees and are of no value to birds as such. Hornbeam is sometimes grown as a hedge mixed with ordinary beech, copper beech and yew, which provides a very pleasing variation of colour. Such a hedge, if 2 metres high and 60 cm. thick, will make an attractive nesting place for large and small garden birds. Hawfinches are very fond of the seeds of hornbeam.

IVY (WALL)

The common ivy seen on old houses and outbuildings has always had rather a bad name for destroying mortar and making a house damp. It is quite true that it does destroy the mortar because it feeds on it, but it will only make a wall damp if the foliage is thin and the rain can penetrate to the wall. If ivy is very thick, it keeps

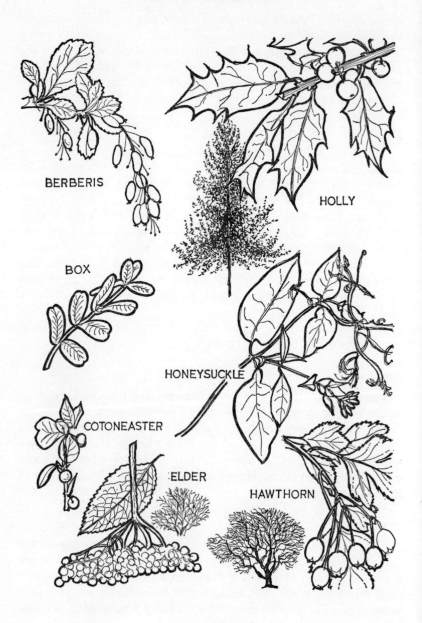

Natural Nesting Sites

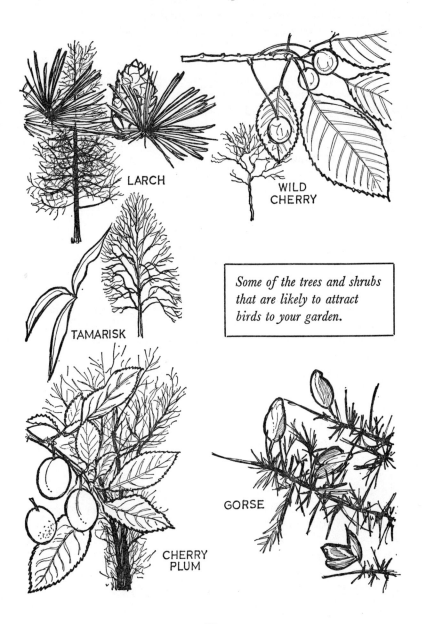

LARCH

WILD CHERRY

TAMARISK

Some of the trees and shrubs that are likely to attract birds to your garden.

CHERRY PLUM

GORSE

the wall dry and well protected from the weather. The natural slope of its leaves acts like a thatch. There is some truth in the old saying 'the ivy keeps the wall up', for if the ivy is removed the outer skin of the brickwork is removed with it, thereby exposing it to the rain and frost which will cause rapid crumbling.

The main value of wall ivy is that it affords such magnificent protection from rain, wind, snow and frost and therefore provides first-class roosting quarters for all garden birds, which are quite safe from predators if the ivy is grown to a height of 2 metres or more; but it is not to be recommended except on a high garden wall or an unwanted outbuilding.

House sparrows, starlings, blackbirds, robins, flycatchers, wrens and wagtails are all fond of wall ivy as a nesting site.

IVY (TREE)

As its name implies, this smaller-leafed type of ivy grows mainly on trees. It has a bush-like structure and provides very good nesting sites and excellent roosting for many birds. It is generally regarded as a slow strangler of trees, but this is a fallacy. Its appearance is attractive and it could well be grown up an unwanted sycamore or ash tree, or a dead apple tree.

The nests of wood pigeons, turtle doves, collared doves, stock doves, mistle thrushes, blackbirds, thrushes, chaffinches, flycatchers, house sparrows and wrens may be found in tree ivy.

JAPONICA (CHAENOMELES)

This shrub can be free-growing or trained against a wall. Unfortunately it is usually grown under the drawing-room window, in which position it requires constant clipping so that the full beauty of its foliage, flowers and fruit can never be seen, and it is too low for nesting birds. It is better to grow it almost free on the lawn, where it will reach a considerable height, will be attractive to look at and will attract birds. It will need occasional pruning to keep its shape.

If grown against a wall it should be allowed to grow at least 2·5 metres high and 30 cm. thick. The strong branch texture makes it ideal for nesting birds and, like the pear to which it is

Natural Nesting Sites

related, it has sharp projections which will deter cats. Its fruit is eaten by blackbirds and starlings.

Blackbirds, thrushes, hedge sparrows, chaffinches and greenfinches will use it for nesting.

JASMINE (SWEET)

This makes a useful nesting site for robins, wrens, blackbirds and thrushes. House sparrows will use it, too. It is very free-growing and requires constant clipping and tying back, but it is easy to fashion nesting sites within its stronger branches at a height of 1·5 metres from the ground.

JASMINE (WINTER)

The yellow-flowered jasmine which always needs a supporting wall, has too weak a growth to be of much value to birds unless it is very old and thick, when robins, wrens, hedge sparrows and wagtails may nest in it.

LABURNUM

This tree which is seen so frequently in gardens, has little value for birds. Its seeds, bark and leaves are poisonous to man, and as a plant it is subject to disease. The sparse, drooping foliage does not provide the right nesting conditions, although goldfinches will occasionally nest at the extremity of one of the upper branches.

LARCH

The larch is a deciduous conifer, growing best in lime soils. Its branches are pendulous or horizontal, and it is therefore not much use to garden birds for nesting. The seeds of the large cones attract tits, greenfinches, nuthatches, redpolls and crossbills. The latter breed locally in England, but are fairly well distributed in the Highlands of Scotland.

LAUREL

It is a pity that this worthy tree has fallen into disrepute and is rarely planted nowadays. Being evergreen it can make an excellent and permanent screen and will stand any amount of cutting. In fact it is difficult to kill a laurel once its roots are well established.

If free-growing, it may reach a height of 5 metres with big, flat, fan-shaped branches which break off easily in a heavy fall of snow. A laurel must be clipped every year to keep it neat, and it is an ideal tree in which to prune out nesting sites.

A laurel trunk, cut horizontally, soon produces shoots around a good nest foundation. The tips of the new shoots should be cut off to promote a thicker growth.

The drawing shows how a laurel tree which has been allowed to grow too big has been cut back with a saw. The cut surface has been made horizontal as a future foundation for the nests of blackbirds and thrushes. Both these birds are fond of laurel for their early nests in April when there is very little leaf cover available. Below the cut surface, new shoots will spring out in the following year, giving excellent concealment for nests. Linnets,

Natural Nesting Sites

greenfinches, bullfinches and hedge sparrows will also nest in laurel. As a roosting site and for protection from heavy storms, a laurel is perhaps the best tree we can have in the garden.

LAUREL, SPOTTED (AUCUBA)

This shrub with its yellow-spotted leaves has a quite different type of twig texture from the common laurel. Its growth is not so vigorous and it is altogether less useful, but if it is kept fairly well pruned a blackbird or thrush may nest in it.

LILAC

Lilac cannot be trimmed in any way as there would then be no blooms, and therefore its branch texture and foliage are too sparse to be of any value to nesting birds except goldfinches and chaffinches. If it is cut back for any reason it will make very strong growth and there will then be nesting possibilities for thrushes and blackbirds.

LIME

When free-growing the lime, like most large trees, is of very limited value to garden birds. It grows very tall and will soon overshadow the rest of the garden, and will usually harbour more greenfly than the birds can eat. A very bushy growth round the trunk can be obtained by pollarding, which will provide nesting places for mistle thrushes, blackbirds and thrushes.

LONICERA (NITIDA)

This is essentially a hedging shrub, and does not have a great deal to commend it as such except that it has a very rapid and thick growth. If it is to be grown to a height useful for nesting birds it will need support against wind and snow. Lonicera is best grown along a protecting wall or fence, the outer side supported by a strand of wire fixed at intervals to the wall. It is subject to 'die-back' of its lower branches.

Smaller garden birds such as hedge sparrows, linnets and warblers may use it for nesting.

MAPLE

Most varieties of maple are large trees with an open type of branch structure which is quite unsuitable for nesting garden birds. The small ornamental trees likewise have the wrong type of branch structure.

MYROBALAN (CHERRY PLUM)

For some reason this very useful tree is not often seen. It is extremely quick growing and with its thorny twigs makes a formidable hedge very quickly. It is very popular with birds and ranks second to hawthorn as the best hedge for nesting sites. Like all hedges, it must not be allowed to get too thick and will require thinning out from time to time. The small yellow plums it produces—which will make an excellent pie—are also appreciated by fruit-eating birds.

Blackbirds, thrushes, linnets, greenfinches, chaffinches, bullfinches, hedge sparrows, yellowhammers, whitethroats and blackcaps will use it for nesting.

MOUNTAIN ASH

Though useless for nesting purposes, this graceful tree is always worth its place in a garden and its berries have a great attraction for mistle thrushes and blackbirds.

OAK

Generally speaking, a large free-growing oak is of no value to nesting garden birds, but it is a tree which attracts a great deal of insect life during the summer and for this reason is useful. Mistle thrushes, wood pigeons and the crow family will nest in an oak tree. Acorns attract tits, woodpeckers, nuthatches, and particularly jays. Squirrels are also very fond of them.

PEAR (see APPLE)

PLANE

These trees are of no value to nesting birds. Wood pigeons nest in them in London squares because they have no alternative.

Natural Nesting Sites

POPLAR

All types of poplar, including Lombardy, are useless as nesting sites for garden birds, their foliage being too thin and their fork structure unsuitable.

POLYGONUM

This rapidly-growing creeper, sometimes called 'mile-a-minute', can be extremely useful for nesting birds if grown against a wall or perhaps over an unsightly shed. It grows into a thick mass in which blackbirds, thrushes, hedge sparrows, linnets, greenfinches, robins and wrens may nest.

PRIVET

It is most unfortunate for garden birds that privet enjoys such wide popularity. From a gardener's point of view it makes the ideal hedge, but from a bird's point of view the neat, well-clipped privet hedge is the last place in the world for a nesting site. The twig structure is too dense for a bird to enter and the height from the ground is usually insufficient.

The only possible way to make such a privet hedge at all useful for nesting birds is to allow it to grow about 15 cm. a year until it is 2 metres high, and to trim out the inside of the hedge periodically. The appearance of the hedge will certainly not be so neat, but such birds as hedge sparrows, linnets and blackbirds may then use it for nesting. An old kettle put inside the hedge may attract a robin.

PYRACANTHA

This tree, which produces such colourful berries, responds well to training up a wall and is therefore more often seen growing against a house. It makes an excellent site for nesting birds and thrushes and is a good place to put an open-fronted box for a robin. Pyracantha has a strong twig texture with some very sharp projections which act as a deterrent to climbing cats. It can also be grown free as a 'standard' up to 3 metres high, and will look very imposing if kept lightly pruned to preserve its shape.

RHODODENDRON

Wherever the soil is acid, rhododendrons should be grown if we wish to encourage birds. Perhaps the greatest asset of the rhododendron is the protection from all types of weather which it gives, when bushes are grown in a mass. It is a good protection against wind because its branches grow right down to the ground and there is no under-draught. In very cold weather frost and snow take a long time to penetrate a good thick clump of rhododendron, which thus provides a valuable feeding ground among the dead leaves underneath.

Rhododendrons thrive on pruning so, when possible, should be pruned back annually after flowering to keep them really bushy. If they are never cut back their growth will become long and straggly and not particularly useful to birds. It may be for this reason that some bird authorities say rhododendrons are not much used by nesting birds; but provided the growth is moderately strong, upright and thick, they will be used by blackbirds, thrushes, hedge sparrows, greenfinches, linnets, lesser redpolls, chiffchaffs, garden warblers and whitethroats. In Northern England and Scotland, chiffchaffs seem to have a particular liking for rhododendron bushes. Being one of the earliest migrants, often before the cold weather has finished, they no doubt seek the protection of dense clumps of rhododendrons where they can find insects.

Rhododendrons respond so well to pruning and training that they can also be grown very successfully against a wall or trellis work, and grown in this way they will be most useful both for nesting and roosting. This applies, of course, to the commoner types of rhododendron which have a vigorous growth. The dwarf varieties are of no use to birds.

ROSE (CLIMBING)

A climbing rose must be grown on trellis work, over an archway, or against the stump of a dead tree, if it is to attract nesting birds. The older and thicker the rose, the better. The chief disadvantage with roses is that the old and strong wood, which is what the bird needs most for nest foundation, must be pruned out if the rose is to look its best.

Natural Nesting Sites

Blackbirds, thrushes, house sparrows, robins and wrens may use it for nesting.

ROSE (HEDGING)

Roses can be grown as a flowering hedge and a good variety is 'Penelope'. It makes a very useful hedge for nesting birds provided it is allowed to grow to about 1·5 metres high and 60 cm. thick.

Linnets, hedge sparrows, bullfinches, greenfinches, whitethroats, blackbirds and thrushes will nest in such a hedge.

SNOWBERRY

This graceful shrub has too flimsy a twig texture to be of any real use to nesting birds unless it is grown in thick clumps and occasionally clipped. Blackcaps, whitethroats, bullfinches and linnets might then nest in it.

SPINDLE

This tree is of no use to birds except for its berries. It attracts blackfly, which insectivorous birds appreciate.

SYCAMORE

From a garden bird's point of view this tree when free-growing is almost useless as it does not provide the right sort of branch structure at a suitable height. Apart from the millions of greenfly which it often attracts, it has no advantage. Moreover, it causes a great deal of work in a garden when its showers of fertile seedlings have to be pulled up. The best thing to do, if the tree is not too large, is to pollard it or cut it right down. If pollarded, the sprouting branches may be used by a mistle thrush or blackbird for nesting. The trunk can be used for tit boxes or artificial nesting holes.

TAMARISK

The foliage of this attractive tree is rather too thin to be useful for birds, but where it is grown in clumps mistle thrushes, blackbirds and thrushes may nest in it.

VIRGINIA CREEPER

One might think that virginia creeper would be ideal for birds but, except in very old vines, there is no solid structure on which a bird can fix a nest. Virginia creeper grows flat against the wall, relying on its suckers to keep it up. If grown over a trellis work, it then provides a well-protected nesting site for blackbirds, house sparrows, thrushes, robins, wrens and flycatchers.

WALNUT

Like most big trees, walnuts have not much to attract garden birds unless they are rotten, in which case woodpeckers may use them for nesting. Rooks, woodpeckers and tits are very fond of the nuts.

WEEDS AND GRASSES

A piece of rough ground covered by long grass, with nettles, groundsel, shepherd's purse, chickweed, dandelions and other weeds, with here and there some brushwood to act as nest support, is of considerable value to garden birds. Those which nest on or near the ground, such as the warbler family, frequently choose an area like this in which to nest. A whitethroat, for example, often builds its nest on old brushwood within a clump of nettles, the latter acting as a deterrent to many predators. Robins like nesting in a high bank under a tuft of dead grass.

Many garden birds use long pieces of dead grass for nest construction and often have difficulty finding it in built-up areas where gardens are kept highly cultivated.

Seeding weeds form an important part of the diet of finches. Insectivorous birds will generally find much more insect life in a rough part of a garden than in a more cultivated part.

Where space allows, therefore, a piece of 'wild' garden is a great asset to the local bird population.

WEEPING WILLOW, WEEPING BEECH

It is very unfortunate that all the attractive varieties of 'weeping' trees are quite useless for nesting birds as the branch structure is unsuitable.

Natural Nesting Sites

WILLOW

Free-growing willow trees of all varieties are of no value to nesting birds, but the pollarded willows seen so often beside rivers develop, as a result of constant cutting back, a thick stumpy growth with numerous holes and crannies. Such a tree can provide many nesting sites for a variety of birds including owls, tits, tree creepers, flycatchers, wild duck and wood pigeons as well as the commoner garden birds. A tree such as this, with so many nesting possibilities, is worth wiring round as a protection against ground predators. Weasels and stoats are particularly attracted to a tree of this sort with all its cavities.

WISTERIA

Wisteria, beautiful though it is when in full bloom, is not of great value to nesting birds. It is a slow growing vine which can be a nuisance if its gets round drainpipes or gutters. Sparrows are particularly fond of the flowering shoots. A very old wisteria which is thick and twisted may be used by blackbirds, thrushes, house sparrows, wagtails and spotted flycatchers for nesting.

YEW

Yew trees, sometimes 400 years old and of huge girth, are often seen in churchyards, where they were no doubt planted to prevent the local population from grazing their animals, which would have been poisoned by eating the foliage. The pulpy part of the yew berry is not poisonous and mistle thrushes, thrushes and blackbirds eat them readily. The seeds of the berry are poisonous and the leaves are a powerful narcotic. Despite these properties, old yew trees contain many cavities in their trunks suitable for tits, starlings, robins, wrens and tree creepers.

Like the box tree, yew was much used in the old topiary style of landscape gardening and was, by patient clipping and trimming, fashioned into very neat hedges and bizarre shapes. Grown in this way, yew is far too thick for nesting birds.

A free-growing yew can be made attractive to nesting birds by cutting back all long, straggly branches to keep the tree in a bushy shape. If the leading end of a long yew branch is cut off, a thick bushy growth will sprout from the sides of the branch. Wrens are particularly fond of such a nesting site.

Another method of obtaining nesting sites, which applies particularly to the free-growing yew because of its long straight branches, is to pull the branches together in the form of a column, by encircling them with twine. The growing ends of the branches should be cut off to increase side sprouting. After about two years this will develop a very thick tree affording good nesting sites for wrens, blackbirds and robins.

In modern gardens, yew trees don't seem to be very popular, presumably because they are rather expensive, rather slow growing and have poisonous qualities; but there is no doubt that yew makes a very good evergreen hedge and a good windbreak, and though it is by no means an ideal hedge from a bird's point of view, it can be fairly useful if kept lightly clipped and allowed to grow about 7 cm. each year until 2 metres high. Grown in this way blackbirds, thrushes, greenfinches and hedge sparrows are likely to nest in it.

CONCLUSION

As a general rule large trees such as the oak, sycamore, beech, lime, elm and ash are of very little use to garden birds for nesting purposes, for the simple reason that they do not provide the right sort of nesting site at the correct height, with the necessary leaf-cover and twig structure. Larger birds like rooks, crows, jays, magpies and wood pigeons will use such trees, but these species seem well able to look after themselves and so do not need encouragement. Big trees, however, particularly oaks, limes and sycamores, do provide a valuable source of insect life, including greenfly, which are much sought after by smaller birds such as warblers and tits.

In a small garden large trees can be a problem. In the spring and summer they take up a great deal of moisture with their spreading roots. Their shade is welcome sometimes on a hot day, but they deprive many surrounding shrubs and flowers of sunlight and stunt their growth. In the autumn the leaves and seedlings of big trees can be a nuisance, while in the winter, snow may break boughs off, creating quite a lot of trouble.

Natural Nesting Sites

One solution which may suit both man and the garden birds is to pollard the tree, provided it is not too large. The growth of the tree is then under control and the sprouting branches can be pruned back every year or two. It will look quite neat when the leaves are on it and will be much more useful to nesting garden birds. Mistle thrushes may choose such a site, as well as blackbirds and thrushes.

As we have seen, it is the trees of medium and small size which are most useful to birds. Some of these are evergreen and others deciduous, and by correct pruning and cutting these can be made into ideal nesting sites.

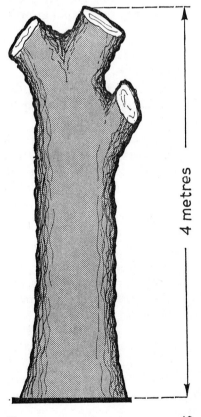

A pollarded tree. Sprouting branches can be pruned back every year or two.

If garden birds had a choice of nesting place from among all the trees listed above, the tree-nesting birds would almost certainly choose the hawthorn. This is certainly the best deciduous tree, while the best evergreen is probably the laurel, which provides weather protection throughout the year and is thus very useful for roosting and nesting. This is particularly ideal for the blackbird and thrush, which often start nesting operations at the end of March when leaf cover is very difficult to find. The rhododendron comes a close second, but its usefulness is limited by the fact that it grows only in a lime-free soil.

Wild blackberry, myrobalan plum, blackthorn, beech hedge and rose hedge all come high on the list. Of the climbing plants tree ivy is perhaps the best all-rounder, while honeysuckle and polygonum are particularly useful owing to their rapid growth.

ONE METHOD OF PLANTING OUT A SMALL, NEW GARDEN TO ATTRACT GARDEN BIRDS.

Area approx. 15 × 30 metres. If there is no fencing, wire netting (large mesh 1 metre width) should be placed round the boundary to exclude cats.

1. *Cultivated blackberry or myrobalan plum hedge. If a screen is required on the north side,* cupressus lawsonii *or* macrocarpa. 2. *Hawthorn hedge on west side ($\frac{2}{3}$ metre wide).* 3. *Rubbish tip and rough patch for seeding weeds or wild blackberry.* 4. *Pruned fruit trees.* 5. *Laurel or rhododendron hedge (at least $\frac{2}{3}$ metre wide) for nesting, roosting and wind protection on the east side.* 6. *Beech hedge ($\frac{2}{3}$ metre wide).* 7. *Rustic archway with climbing rose, honeysuckle, clematis or polygonum.* 8. *Pathway.* 9. *Shed with creeper, climbing rose or polygonum on trellis work. Rain water from the shed roof can be piped under the path to run into the pond.* 10. *Garage on which tit boxes, open nest boxes, bundles of twigs or creepers on trellis can be fixed.* 11. *House. Clematis, climbing rose, white jasmine, virginia creeper, ceanothus, apple, pear, plum or peach can be grown on trellis work against the house wall. Pyracantha, rhododendron and japonica, though they will stand on their own, are more easily trained and more useful to birds if grown against trellis work.* 12. *Road hedge. Yew, beech, hedging rose or* cupressus lawsonii.

Natural Nesting Sites

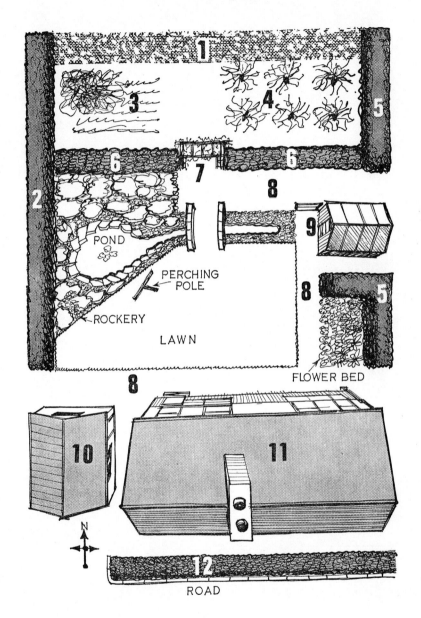

CHAPTER THREE

Artificial Nesting Sites

WHEREVER THERE is a shortage of natural nesting sites we must try to provide an alternative. This can be done for quite a number of birds if we know what is required.

Some tree-nesting birds, particularly the blackbird, have already adapted themselves to building in open nest boxes, in garden sheds, on drainpipes and in many odd places in close proximity to man. This shows a trend in bird behaviour of which we must take full advantage.

Generally speaking, a nesting site against the wall of a house can be made one of the safest places for a bird to rear her young. The presence of man automatically helps to keep away a lot of the nesting bird's greatest enemies, and if the artificial nesting device is correctly placed the risk from cats, rats and mice can be reduced to a minimum. The house wall is one of the best places in any garden on which to fix nesting devices, and moreover it may be the only place in some of the new housing estates where large dividing hedges are not allowed.

NEST BOXES FOR BLUE TITS, GREAT TITS,
COAL TITS AND MARSH TITS

1. Construction

Nesting boxes can be made in many different ways. They can be cut out of an old log, they can be made of the most expensive

The Bird Gardener's Book

wood and finely finished or they can be put together in an amateurish way with old bits of wood. All will be used by tits provided the conditions given below are fulfilled.

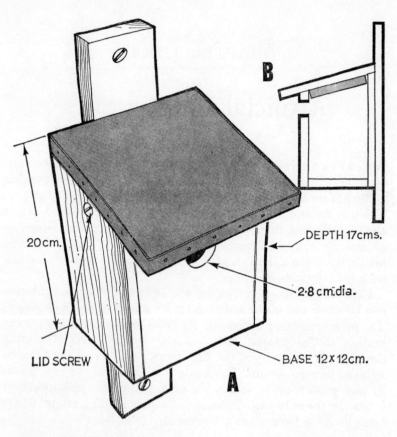

Nest box for blue tits, great tits, coal tits and marsh tits. A. Front view. B. Cross-section showing roof reinforcement (shaded area).

The construction should be on the same lines as shown in **A** above, using wood not less than 2 cm. thick. The size of the entrance hole is most important. It must not be more than 2·8 cm.

Artificial Nesting Sites

diameter (the size of a 10-pence piece) for blue tits, coal tits and marsh tits and 3 cm. for great tits. The entrance hole should be placed about 1·5 cm. from the top and need not be central. A perch under the hole is not required, in fact it is a disadvantage because it gives sparrows an opportunity to sit on the perch and annoy the occupant. Tits are so acrobatic that they have no difficulty at all in flying straight up to the nesting hole and clinging to the sides of it before entering. The roof of the tit box should overhang the entrance hole by about 5 cm. This overhang keeps out the weather and makes it much more difficult for squirrels or cats to attack the occupant from above or for field mice to climb into the box.

The roof of the box must be removable in order to clear out the old nest. For sanitary reasons, tits do not like breeding on top of an old nest in which there have been young birds. Hinges are generally not satisfactory as the roof may be subject to slight movement in a high wind. The best way to get a snug fitting is to fix a piece of wood on the under-surface of the roof in such a way that it fits into the top of the box (see shaded part in **B**), thus providing something into which a screw can be fixed on each side. If the screws are oiled before they are put in, they can easily be removed each year for cleaning out the box.

Another advantage of using screws instead of hinges is that it prevents people from having a well-intended peep to see 'how things are getting on'. Such interference is never appreciated by the bird and she may desert her eggs unless incubation is well under way and the eggs are near to hatching. This rule applies to all garden birds. The less they are disturbed, the better.

Most birds have no sense of smell, so it does not matter if the box smells of creosote or wood preservative. Waterproof felt should always be put on the roof and over any crack which might let in a draught.

2. *Positioning*

The position of the box is the next important thing. It must always be placed out of the reach of cats, rats, mice and squirrels. As none of the latter can climb a well-built brick wall, the walls of a house or building are generally the safest place and should be considered first. The box should be placed at a height of about 2 metres; thus, if it is placed close under the eaves of a garage, outbuilding or bungalow it will be in an ideal position, with protection from above and below.

It is hardly necessary to say that the box should be fixed very firmly so that there is no movement in a wind. This is best done by hammering a fixing board to the back of the box before putting it up, which makes nailing to the wall much easier, and also prevents damp getting into the back.

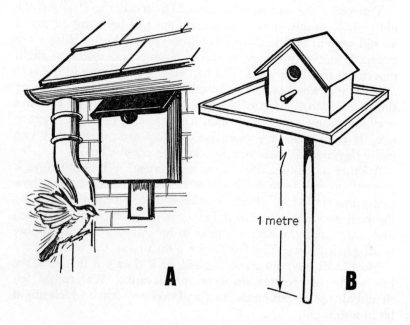

How nest boxes should and should not be placed. A. The box is ideally placed under the eaves of an outhouse at the correct height. B. A combination tit box and bird table which has everything wrong with it. The box is too big and the hole is too big. It has been placed on a feeding table in an open place at a height which would make it

Artificial Nesting Sites

If the box is sheltered under the eaves of a roof, it does not matter if it faces north, south, east or west.

When fixing tit boxes on trees, it is always best to choose the trunk of a large tree with smooth bark. An underslope of the trunk, or a large branch coming off the trunk is better still, for this makes it more inaccessible to climbing predators, and the tits certainly won't worry about their box being tilted at an angle.

Tit boxes on trees should never have any branches near the entrance hole. Tits like an unobstructed view when entering and leaving their nest, and do not like perches for predators just outside the nest. Weasels and field mice, for example, can more easily

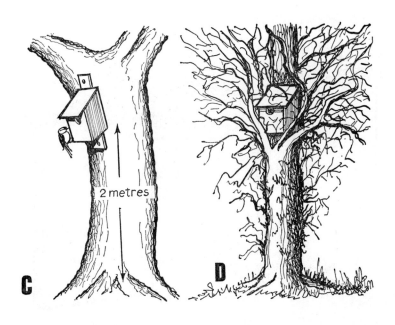

very easy for a cat to spring up on to the table. C. *The box has been well placed at the correct height on the underslope of a large tree.* D. *A good box badly placed. It will be accessible to cats and field mice, and the foliage will obstruct the view round the entrance hole.*

gain access to a tit box if branches grow across the entrance hole. This is a very important point—in fact, it is really a waste of time putting up a tit box surrounded by branches.

Boxes on trees should not be placed facing south into the full sun but should face west, or north-west if the position is exposed. There is, however, no hard and fast rule about this, and it depends a great deal on shelter provided by nearby trees.

Generally speaking, remember when fixing tit boxes that there should never be anything between the bottom of the box and the ground, and no foliage in front of it. Once the box has been

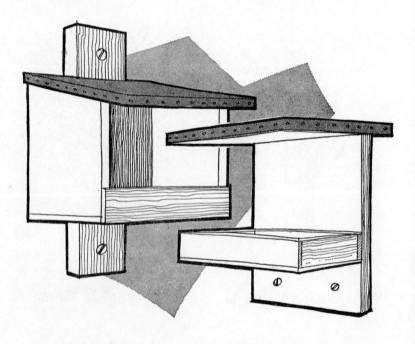

Open-fronted nest box for robins, wrens and flycatchers; and box with open front and sides for blackbirds and thrushes.

Artificial Nesting Sites

successfully occupied, don't move it, because it will almost certainly be used every year by the original occupants or their progeny. If the box is not occupied, its site should be changed.

NEST BOXES WITH OPEN FRONTS FOR ROBINS, WRENS AND FLYCATCHERS

These boxes are quite simple to make and the old nest can be easily removed through the front.

For robins and wrens the box should be placed about 1·5 metres from the ground in a thick creeper. If it is intended for a flycatcher it should be placed in a fairly exposed position on a wall.

NEST BOXES WITH OPEN FRONT AND SIDES FOR BLACKBIRDS AND THRUSHES

This box is wider than the open-fronted box in order to accommodate bigger birds such as blackbirds and thrushes. These birds like to be able to see over the sides of the nest, and to have freedom of movement in case of danger.

The siting of such boxes is important. They should be about 2 metres from the ground. If put on an open wall they must be partly hidden by a bundle of branches above and below and fixed rigidly. If fixed on to the wall of a gabled house or garage, they should be placed below the eaves, preferably with a few evergreen branches around them.

ARTIFICIAL HOLES IN TREES

An artificially-made hole in a dead or partly-dead tree is a useful alternative to a nesting box, and although there is no evidence that such holes are preferred to nest boxes—indeed, it may be the other way round, for great tits and pied flycatchers will sometimes leave natural holes if boxes are available—it is always interesting to see how different birds react to the different accommodation offered.

A tree-hole should be about 10 cm. deep, 10 cm. wide, and about 15 cm. high. If it is to be used by tits, the entrance hole in the front must be made the appropriate size to exclude sparrows.

After chipping out the hole, a front should be fitted, using wood about 2 cm. thick. The front can be secured in position by a vertical strip of wood, as shown, leaving the nails half hammered in so that the front can be removed easily for cleaning purposes. The hole should be about 2 metres from the ground, preferably on the underslope of a tree to make it more difficult for cats and squirrels to interfere.

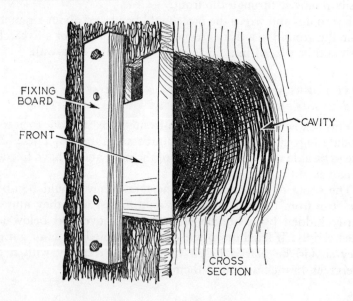

Artificial nest hole chipped out of a tree.

Nuthatches may use such a nesting hole, as they will also use ordinary tit boxes. If the hole is too small they can enlarge it, and if too big they will plaster it up with mud to keep out unwelcome visitors. They also plaster up any cracks in the nest box, including the inside of the lid, which is an important point to remember for anyone inspecting the nest box. Strangely enough, although the nuthatch has a very powerful beak, it seems unable to cope with invasion by sparrows.

Artificial Nesting Sites

An open artificial nesting hole can also be very popular with flycatchers. In such a case no fitted front is required but only a piece of wood 2·5 cm. wide, nailed across the bottom of the opening horizontally. Flycatchers need only a shallow nesting site.

HOLES IN THE WALLS

A suitable hole in a brick wall is made by removing one brick completely and splitting it lengthways, leaving a thickness of about 4 cm. Clean out the cavity of loose mortar and then lightly cement the split brick back into place, with a gap about 2·5 cm. at one end of the brick as an opening. The nest cavity can easily be cleaned out each year by tapping out the brick. The nest hole should be not less than 1·5 metres from the ground. A piece of slate may be inserted above the movable front brick as a support for the bricks above. It sounds a tedious process, but it will be

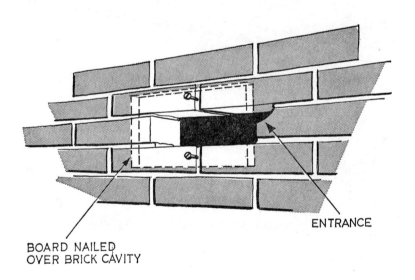

BOARD NAILED
OVER BRICK CAVITY

ENTRANCE

A nesting site made by a brick cavity in a wall.

very worthwhile and will provide a safe and permanent nesting site for tits.

It is equally effective but less pleasing in appearance to hammer a board over the cavity, chipping a corner off the brick next to the cavity to act as an entrance hole.

NESTS IN SHEDS

To encourage swallows to nest in outbuildings, a small hole about 5 cm. × 7·5 cm. wide should be made at the top of a door. The door can then be closed and the swallows can get in and out as they please. A 40 cm. length of 5 cm. × 5 cm. wood should be hammered up on the highest place in the outbuilding in a position

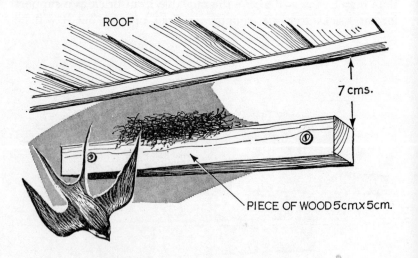

A nesting site for swallows in an outbuilding.

where cats, mice or rats cannot climb. It is wise to hang a piece of sacking directly under the nest (like a circus net) as the young swallows will make a very considerable mess on whatever is below. A similar support can be fixed up in a house porch, provided this does not have a transparent roof, which swallows dislike when nesting.

It is a popular belief that swallows breed under the eaves of houses. This is incorrect. It is the house martin which breeds under eaves. Swallows nearly always build their shallow, cup-shaped nests of mud on some sort of support or ledge such as a cross-beam inside a barn, large shed or garage. The swallow's nest, unlike that of the house martin, is quite open, so the swallow always likes a good roof over its head to protect the nest from winged predators.

OPEN SHEDS

If there is an 'open' shed or a lean-to structure against a wall in the garden, this can always be made into a potential nesting place for blackbirds and robins if a shelf is constructed in the highest part beyond the reach of cats, rats and mice.

NESTS ON TRELLIS WORK

The trellis work which we commonly see on houses supporting clematis and other climbing plants is unfortunately of little use to nesting garden birds because it is usually too flimsy and is always fixed too close to the wall. A bird such as a blackbird cannot find enough nest support on such a trellis unless there happens to be a very thick creeper growing up it.

This lack of nest support can be overcome fairly well on an existing trellis by tying a length of wood 5 cm. × 5 cm. wide against the trellis at a height of about 2 metres. This will give the necessary support for a nest.

When a new trellis work is made and it is hoped to encourage nesting birds, the trellis should be made in a certain way—it should be square, not diagonal. The horizontal parts should be made of 5 cm. × 5 cm. lengths of wood, as it is these which provide most of the support for the fairly heavy nest of a blackbird or thrush. A robin or wagtail can get nearly all its nest on a width of 5 cm. The rest of the nest support is derived from whatever shrub or creeper is growing up the trellis. The vertical strips need only be 2·5 cm. thick. The trellis is best fixed to the wall with angle irons.

Trellis work need not necessarily cover a large area, and it is better if it does not reach to the ground. Six or eight squares of trellis behind a climbing rose, at a height of 2 or 3 metres, will be good both for the rose and for nesting birds as well as giving a much neater appearance to the wall. There are many places on a house where an ugly wall can be made quite attractive by using trellis work. The shrubs recommended are honeysuckle, white jasmine, japonica, forsythia (golden bell), clematis (but not types that require annual pruning), polygonum, virginia creeper, rhododendron, climbing rose and even a domestic blackberry.

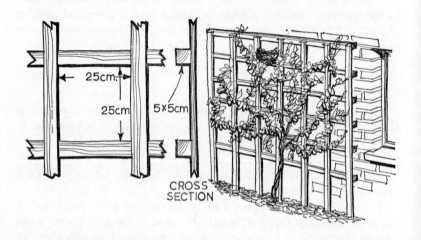

Trellis work suitable for supporting large and small nests.

NESTS ON ARCHES

A garden always looks better if it contains one or two rustic arches. Not only do arches 'break up' the garden and make it visually more interesting, they also provide a floral display at a higher level and are a great advantage where space is short.

Artificial Nesting Sites

A rustic arch suitable for nesting birds should be about 2 to 3 metres high and measure about 75 cm. between horizontal poles. It is a good plan to put 2·5 cm. wire mesh netting over the top of the arch, for this will act as a support for the climbing plant and keep the underside of the arch clear, as well as forming a good foundation for any nest. To increase the leaf cover it is worth growing over the arch something like the clematis 'Montana' intermingled with, say, a climbing rose.

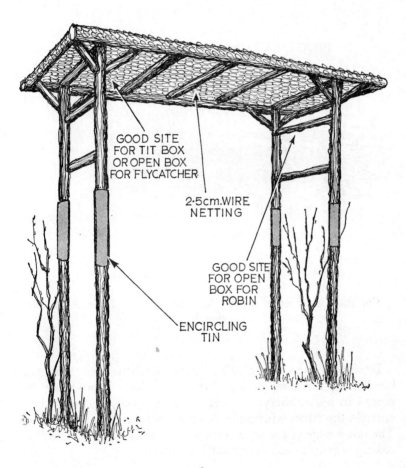

BUNDLES OF TWIGS

One method of forming a nesting site which should be particularly attractive to blackbirds and thrushes is to wire a bundle of pea sticks together and fix it horizontally at a height of about 2 metres under a protective shelf or under an overhanging roof. It may also be fixed on the wall of a house, but then a protective board must be put above it to hide the nest from above and to keep off rain and sun. Another good place is under a window box on the first floor. In this case a protective board is not necessary.

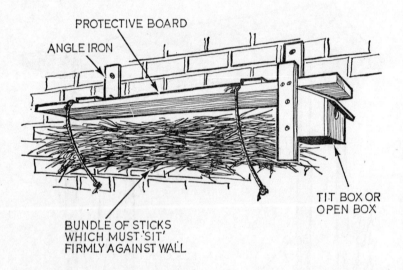

Pea sticks fixed under a protective shelf to attract blackbirds and thrushes.

Bundles of twigs may also be tied vertically to a stake to attract blackbirds, thrushes, robins and wrens. Such bundles are best placed in a shrubbery. If there are cats about, it is essential to encircle the sticks with some large mesh wire netting, as shown. The outer edge of the wire netting must be flexible—cats will not risk a jump on to anything that is not firm.

Artificial Nesting Sites

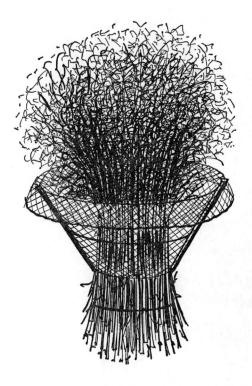

Bundle of twigs with an anti-cat barrier of wire netting.

ARTIFICIAL SITES FOR TREE CREEPERS

Tree creepers usually nest behind a fold of bark which protrudes away from the parent tree, or in cracks and crevices in tree trunks. Artificial nesting sites, made of bark, usually attract predators, and nest boxes of the type illustrated are generally safer for the occupants.

A log of wood about 20 cm. long and 12 cm. in diameter, with rough bark such as elm or pine, should be cut lengthways down the

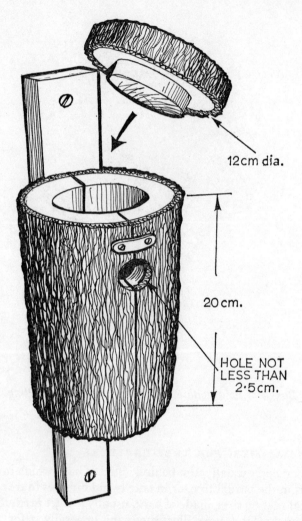

A rustic nest box for tree creepers.

Artificial Nesting Sites

middle, a slice for the lid having been cut off first. The cavity can then be chiselled out from each half. The depth of the cavity should be about 12 cm. and the diameter 7 cm. It is better to carve out what will be the bottom of the cavity rather than replace it afterwards with a different piece of wood. The entrance hole should be oval-shaped and not wider than 2·5 cm.

The box must be firmly fixed on the underslope of a tree trunk or large bough, in a fairly secluded place.

A kettle will make a safe site for a robin's nest.

KETTLES FOR ROBINS

Robins do not nest in the branches of a tree, but usually select some ivy-covered wall, bank or thick creeper. They are thus subject to interference by field mice, which are perhaps their biggest enemy, and it may be for this reason that they are very fond of breeding in kettles. Field mice cannot climb in if the kettle is properly placed so that there are no twigs round the opening.

A kettle fixed firmly to a board under the eaves of an outhouse is very likely to attract a robin.

SPECIAL SITES FOR WRENS

Wrens are particularly fond of sacking as an artificial support for a nest. It is easiest to use a sack, folded so that it is fairly thick, and then hammered firmly to the under-side of a leaning tree trunk, or to the under-side of a bough 2 metres from the ground. It can also be fixed under the eaves of a garage or shed.

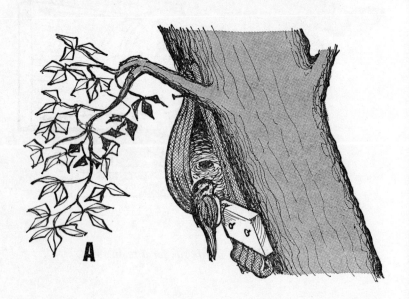

Nesting sites for wrens. Sacking can be fixed to the trunk of a tree (A) or to the underside of a branch (B).

Artificial Nesting Sites

NEST BOXES FOR OWLS AND KESTRELS

If there is a large tree in the garden, it may be possible to attract owls, kestrels and stock doves by putting up a nest box made like a single-holed dove-cote. A small barrel (about 12 cm. diameter) with a hole in one end would also do very well. It should be accessible by ladder in case a grey squirrel decides to occupy it, either as a winter residence or as a nesting box, and should be fixed at least 6 metres high in the tree.

If an owl takes a fancy to the box it may use it both for nesting and as a permanent sleeping place during daylight hours. Jackdaws will also use this type of box quite readily, but it is not recommended that the latter should be encouraged too much as they can be a nuisance on a bird table.

One such box, which was put up for a kestrel, was soon occupied by starlings and, later in the year, by stock doves. The starlings still visit the box regularly even though the stock doves are in full occupation.

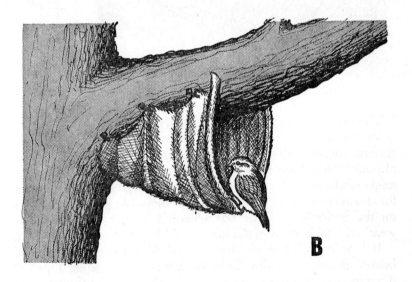

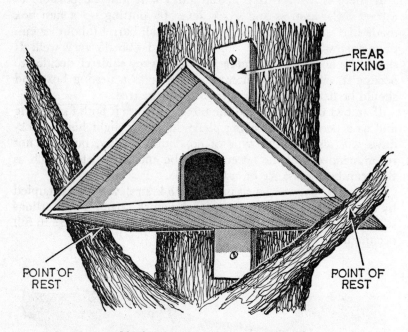

Nesting box for owls and kestrels.

The chimney box, designed by H. N. Southern especially for tawny owls, is a simple structure about 80 cm. long and 20 cm. square. Several holes should be bored in the bottom for drainage, although very little rain will enter the box if it is fixed at an angle of about 30° on the underslope of a large branch. To allow for cleaning out and inspection, a door should be made, preferably on the underslope of the box where it is least exposed to the weather. The bottom of the box should be covered with sawdust.

It is worth while encircling the bough of the tree above and below the box with shiny linoleum at least 60 cm. wide. This will discourage grey squirrels who, once they have found the box, will regularly destroy the eggs or young of any bird ocucpying it.

Artificial Nesting Sites

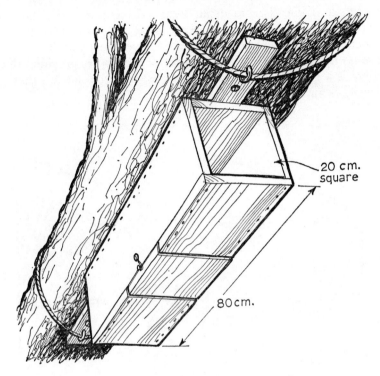

Chimney-shaped nest box for tawny owls.

NESTING SITES FOR HOUSE MARTINS

The house martin builds a nest of mud which is completely enclosed except for a small entry hole through which the bird can squeeze. It always builds fairly high up under the eaves of houses and does not seem particular about locality, for it will often choose a house in the middle of a busy street, provided mud and water for nest construction are near by.

Artificial plastic nests for house martins can be bought* and fixed under the eaves. Such devices certainly attract the birds,

---

* From Clent House Gardens, Clent, Worcestershire.

and if they do not actually use them for nesting they may well build alongside. The artificial nests are supplied with boards fixed above and behind the nest. It is best to put at least two of them side by side where house martins have already built. If there are no martins' nests, try any aspect of the house, provided that there is a good roof overhang. The nest must be not less than 6 metres from the ground.

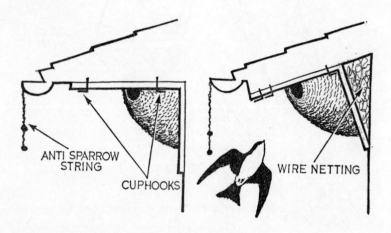

Artificial nests for house martins, one fixed to a flat surface under the eaves, the other fixed beneath a sloping roof.

The drawing shows a flat surface under the eaves, where fixing with right-angled cup hooks is quite easy. This type of hook is useful because it is rust-proof and the nest can easily be detached by twisting the hook.

Where the undersurface of the eaves is sloping, the fixing is not so simple as the nest must be put at an angle, leaving a space behind which should be filled in with wire netting to stop sparrows building.

It is important to remember that sparrows will break into a

Artificial Nesting Sites

house martin's nest by nibbling away the mud and will evict the eggs or young—sometimes, it seems, for the sheer fun of it and sometimes because they want the nest for themselves. For this reason alone it is a wise thing to discourage house sparrows, and a good way to do this is to suspend a series of small weights from the water gutter on pieces of string about 30 cm. long and 10 cm. apart. These will stop the direct approach of sparrows but will allow the martins to sweep in and out underneath.

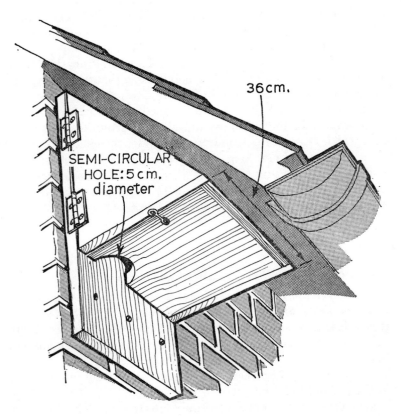

A nesting box for a swift.

NESTING HOLES FOR SWIFTS

These fascinating birds, which spend almost their whole lives on the wing, cannot perch but can only cling to a rough wall long enough to work their way into a nest hole. In old houses, they invariably breed where there are small holes, about 3 cm. in diameter, under the eaves, usually between the fascia board and the top of the house wall. Anyone who has sufficient energy to make a hole of this nature may soon find a swift in occupation. They are comparatively common birds, but spend so much time high up in the air that we rarely see them. They breed freely all over the country wherever there is old property. They can even be seen in Islington, St. John's Wood, near the Edgware Road and in many other places in the heart of London.

The increased use of concrete, glass and metals for new buildings, as opposed to brick and wood, makes it likely that swifts are going to have increasing difficulty in finding nesting holes. Nest boxes have been used and they should be fixed under the eaves in the same way as for the house martin. The important features of the box, which must be made to fit snugly under the eaves, are that it must be about 36 cm. long and the entrance hole must be either underneath or at the side. The entrance hole should measure approximately 5 cm. × 2·5 cm. A small coil of straw placed at the end of the box will provide an artificial nest cup.

PROTECTION OF NESTS FROM PREDATORS

Whenever there is a possibility of any type of nest box or nest in a tree being attacked by cats, squirrels, mice, rats, weasels or stoats, it is a good plan to encircle the tree with a collar of wire netting of the smallest possible mesh. This is not an easy job on a big tree, and it needs quite a lot of wire netting, but it is well worth while. Smaller trees which have or are likely to have nests in them should also be wired.

It is not necessary to hammer nails into the tree. The inner part of the wire netting can be held to the trunk by an encircling strand of wire, while the outer edge can be held up by wire or string

Artificial Nesting Sites

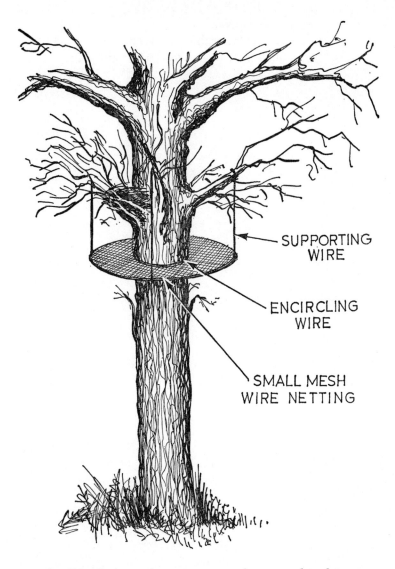

A collar of wire netting protects nests from ground predators.

attached to the branches above. Nothing except perhaps a field mouse can pass such a barrier.

In order to prevent cats climbing an arch, it is a good idea to hammer a length of tin round each of the upright poles.

CHAPTER FOUR

Nesting Preferences of Various Birds

IN A GARDEN the choice of trees and shrubs is usually made to suit the needs of man rather than those of birds, thereby limiting the selection of nesting sites from which a bird can choose. A fuller description, therefore, is given below of the types of nesting site a garden bird would select if allowed a freer choice. It relates to what has already been mentioned under the list of trees and shrubs, and is intended to provide an easier method of reference.

BLACKBIRD

Blackbirds will nest in all evergreens, such as laurel, yew, tree ivy, wall ivy, rhododendron, holly, young spruce and other conifers with suitable branch structure, and occasionally privet. Deciduous trees and shrubs such as hawthorn, blackberry, blackthorn, forsythia, flowering currant, escallonia hedges, japonica, myrobalan plum, honeysuckle, polygonum, pyracantha, are also popular places.

The blackbird, more readily than the thrush or mistle thrush, will build in artificial nesting devices, on trellis work covered by climbing shrubs and, particularly when there are winged predators about, in open garden sheds.

BLACKCAP

This bird likes to nest in small bushes, especially elder. It also likes the horizontal branches of hollies and will nest in ivy and sometimes in low bushes of laurel and rhododendron.

BLUE TIT

The blue tit, probably the commonest member of the tit family, nests in a natural hole of some sort, either in a tree or wall or in a nest box. It will sometimes nest in a street lamp, pillar box, or an old pump. Blue tits usually take much more readily to artificial nest boxes than other tits, but in districts where there are many natural stone walls they often seem to prefer holes in walls to nest boxes, even though they are then far more exposed to such predators as weasels and mice. I recall an instance where a blue tit had built in a rough sandstone wall and showed no interest whatsoever in a tit box close by. Ten days after the young had hatched, the nest was attacked and the young devoured.

BULLFINCH

The bullfinch will nest in both evergreen and deciduous hedges and shrubs. It likes dense foliage, and beech hedges are particularly popular.

CHAFFINCH

The chaffinch usually builds in a deciduous tree such as a hawthorn, blackthorn, or tall blackberry bush. It will also nest in elderberry, tree ivy, climbing rose, japonica, and fruit trees.

CHIFFCHAFF

Patches of thick weeds in which there is twig support, heather, hedge bottoms, and low bushes are chosen by the chiffchaff for nesting.

COAL TIT

This bird nests in holes in trees, in walls and even holes in the ground. Only occasionally will it occupy a nest box.

Nesting Preferences of Various Birds

COLLARED DOVE

Evergreen trees such as holly, spruce and other fir trees, tree ivy, and large deciduous trees where there is good leaf cover, are favourite nesting sites.

CUCKOO

The nest of a hedge sparrow, pied wagtail or robin, is usually chosen to receive the cuckoo's egg.

GARDEN WARBLER

This bird nests in low bushes or blackberry patches where there are nettles and weeds intermixed.

GOLDEN-CRESTED WREN

The under-surface of a branch of a coniferous tree is chosen by this little bird from which to sling its hammock-like nest. It will also nest in ivy, in broad-leafed woodland trees and in gorse bushes.

GOLDFINCH

The goldfinch builds high in the branches of a large tree such as an oak, sycamore or chestnut. Also, much lower, in apple trees, lilac, hawthorn, climbing roses, clematis, tall blackberry bushes and conifers.

GREAT SPOTTED WOODPECKER

This bird usually bores out its own nest hole in a decayed part of a large tree, often an elm, walnut, ash or birch. It has nested in boxes and frequently 'attacks' them.

GREAT TIT

This tit nests in holes in trees, walls, in any suitable cavity and in tit boxes. Upturned flower pots are quite often used. In some areas the great tit population moves entirely into nest boxes.

GREENFINCH

The greenfinch likes to nest in hawthorn, blackberry or blackthorn hedges, certain conifers such as *cupressus lawsonii* and *macrocarpa*, laurel, rhododendron, holly, myrobalan plum, climbing roses on an arch and in yew.

GREEN WOODPECKER

The decaying part of a large tree, such as an elm, walnut or birch, is always chosen by this bird for its nest hole.

GULLS

In coastal areas, certain species of seagulls must be added to the list of garden birds. Those which are most frequently seen are the herring gull, common gull and black-headed gull. The herring gull, the largest and probably the least welcome visitor to a garden where there is a feeding table for smaller birds, has now adapted itself to nesting on buildings as well as on cliffs. The other two species nest in colonies in marsh land and sandhills and therefore visit gardens mainly during the winter months.

HAWFINCH

Though comparatively rare in gardens, the hawfinch will nest in tall fruit trees, in evergreens and in certain coniferous trees.

HEDGE SPARROW

Any low, thick bush or hedge, whether evergreen or deciduous, will be chosen as a nesting site.

HOUSE MARTIN

These birds nest under the eaves of a roof and never less than 6 metres or so from the ground.

HOUSE SPARROW

The house sparrow nests in holes in any type of building, in drainpipes, ventilators, house martins' nests, tree holes, wall ivy, and in nest boxes. It will build its own nest, if necessary, high up in any garden tree.

JACKDAW

This bird is a cavity nester and builds in holes in trees, in cliffs and in old ruins. Chimney pots are a favourite place and it will also use nest boxes and dove-cotes.

JAY

Favourite nesting places of the jay include conifers, such as tall spruces, tall hawthorn trees or any big tree where the leaf cover is dense.

KESTREL

Large cavities in old trees, old carrion crows' nests and ledges in church towers and other tall buildings, are often chosen by kestrels for nesting. Nest boxes, if placed high enough, may also be used.

KINGFISHER

The kingfisher often appears in gardens near rivers or lakes. It always nests in a tunnel in a vertical bank usually overlooking and never very far from water.

LESSER REDPOLL

This little bird nests in fairly tall evergreens such as rhododendron and in deciduous trees, particularly when they are grown close together in a shrubbery. It is seen mainly in northern districts.

LESSER SPOTTED WOODPECKER

The nesting habits of this bird are similar to those of the green woodpecker.

LESSER WHITETHROAT

Thick, overgrown hedges, well entangled with blackberries, nettles and other weeds, are a favourite nesting place.

LINNET

The linnet is particularly fond of gorse, but will nest in thick evergreens such as laurel, lonicera and box. Hawthorn, blackberry bushes, garden hedges and weedy thickets are also popular.

LONG-TAILED TIT

The nest is built in a tall blackberry bush, blackthorn tree, hawthorn hedge or gorse bush. The long-tailed tit invariably selects a thorny tree to protect its nest.

MAGPIE

The favourite garden nesting site of the magpie is a tall pear tree. It will also nest in an elm, oak, ash, sycamore, beech, Scots fir and in tall hawthorn trees or hedges.

MALLARD

The mallard or wild duck usually nests in thick cover on the ground near water. Sometimes it nests above the ground in a pollarded willow or tree stump, occasionally on a building some distance from water. Mallards take quite readily to raised nesting boxes, in which they are much safer from magpies and rats.

MARSH TIT

This bird nests mainly in naturally-occurring holes in trees. It occasionally uses a nest box.

MISTLE THRUSH

The mistle thrush often builds in an exposed fork of a large tree such as an ash, beech, cherry, apple or pear. Where magpies abound, it will choose an evergreen such as a tall rhododendron, laurel, yew or holly.

Nesting Preferences of Various Birds

MOORHEN

The favourite nesting site of a moorhen is on a half-submerged branch which has fallen from an overhanging tree into the water. It will also nest in rushes and in thick foliage on banks, usually beside still water.

NIGHTINGALE

Very thick undergrowth consisting of bramble, bracken, sprouting hazel nut stubs, grass and weeds, or the bottom of an overgrown hedge, are favourite nesting places of the nightingale.

NUTHATCH

The nuthatch nests in a naturally-occurring tree-hole or in one made by a woodpecker. It neatly plasters up the entrance hole with mud to the exact size it needs. It will occasionally breed in nest boxes.

OWL, BARN

As its name suggests, the barn owl prefers barns and old buildings for nesting purposes, where it nests on a flat surface such as a beam. It will also nest in the large hollow cavities often found in elms. It will sometimes occupy a nest box if this is suitably positioned in a quiet place, high in a tree or ivy-covered wall.

OWL, TAWNY

The tawny or brown owl nests in tree-holes and in cavities in buildings. It sometimes nests in artificially-made nesting boxes.

OWL, LITTLE

Naturally-occurring holes in trees, such as in elms and pollarded willow trees, are the favourite nesting site of the little owl. It will also occupy nest boxes.

PARTRIDGE

Though not considered as a garden bird, the partridge is particularly fond of coming into a garden from a neighbouring field to nest quite close to the house in a flower border, in long grass or under a garden hedge.

PHEASANT

Like the partridge, pheasants will come into a garden to nest, using man as protection against some of the sitting pheasant's many enemies.

PIED FLYCATCHER

The pied flycatcher, found mainly in the northern counties of England and Wales and the southern part of Scotland, usually builds in a hole in a tree but occasionally will use a cavity in a building. It is very much a woodland bird and in certain Forestry Commission areas it breeds freely in the nest boxes provided.

PIED WAGTAIL

This charming bird usually nests in ivy-covered walls and thick foliage overlooking a river or lake. It will occasionally nest in places such as a farm where there is a limited supply of open water but a plentiful supply of flies.

RED-BACKED SHRIKE

The red-backed shrike almost always builds in a thick thorn bush or in a mass of brambles. Roadside hedges are its favourite habitat.

RING OUZEL

This bird might be called the blackbird of the moorland. It nests in the stone walls and heather banks of open country.

ROBIN

The robin nests in thick ivy on a wall, in creepers growing on trellis work, in steep, ivy-covered banks such as are seen on the roadside, in naturally-occurring clefts in trees and in very thick cypress trees. It also nests in open nest boxes or receptacles such as kettles, concealed in a thick creeper against a wall or in a hedge.

Nesting Preferences of Various Birds

ROOK

Rooks always nest in colonies, often choosing a clump of large trees in a town garden. Beech, elm, ash, sycamore, oak and chestnut are most frequently used.

SAND MARTIN

A vertical sandy bank is always chosen, not necessarily near water. Old sand quarries are favourite places. Anyone who has a sandy, sloping garden could develop a suitable nesting site, but the vertical face of the bank must be about 2 metres high as a protection against predators.

SISKIN

Though breeding only locally in Scotland and Ireland, the siskin often appears in gardens in the winter. It usually nests in conifers.

SPOTTED FLYCATCHER

The flycatcher builds mainly on masonry and in niches in walls. It will also build on the stems of tree ivy and virginia creeper and in shallow tree cavities. Open nest boxes, fixed against a downpipe on a house wall at a height of 2 to 3 metres, may well be occupied by a flycatcher.

STARLING

This bird is a cavity nester and will nest in holes under the roof guttering, in chimneys and in any suitable crack or cranny in a building. It will nest in tree-holes, sometimes evicting the occupant; also in sea-cliffs and quarries. Nest boxes with a big enough entrance hole will invariably be occupied by a starling, if placed fairly high up.

STOCK DOVE

The favourite nesting site of the stock dove is in thick ivy either on trees or on a wall, in natural holes in trees and in nest boxes.

It also likes large lime trees which have a very thick and bushy outgrowth round the main trunk.

SWALLOW

The swallow will always build under a roof of some sort. Tractor sheds and barns are favourite places. It will nest in porches and in garages provided they are high enough. The nest is almost always on a wooden beam or ledge, but a swallow can build a nest on a flat brick wall. Such nests are liable to collapse as the young get heavier.

SWIFT

A small hole under the eaves of an old building or a gap between the bricks or stones in a church or ruin is usually chosen.

THRUSH

The thrush has very similar nesting preferences to those of the blackbird, its close relative. Its first nest will almost certainly be built in an evergreen, such as laurel. The thrush is perhaps less inclined to use nest boxes than the blackbird.

TREE CREEPER

This bird builds in vertical crevices such as those found when loose bark splits away from a tree trunk. Also, it might nest behind a loose board in a barn or shed. It will breed in nest boxes.

TREE SPARROW

The tree sparrow, which is very similar to the house sparrow in appearance, will nest in tree-holes, cracks and cavities in disused buildings and in nest boxes. It is very fond of nesting in thatched roofs, otherwise it usually keeps away from houses.

TURTLE DOVE

Hawthorn, spruce, tree ivy, hazel trees and almost any tall hedge with good leaf cover are favourite nesting sites of the turtle dove.

Nesting Preferences of Various Birds

WHITETHROAT

This bird nests in thick hedgerows, in blackberry bushes, and in thick masses of weeds, especially nettles and low scrub.

WILLOW WREN

The willow wren usually builds its dome-shaped nest on a piece of rough ground where leaves and sticks have collected. A favourite spot is under the low overhanging branches of a tree in a tuft of dead grass.

WOOD PIGEON

The wood pigeon nests in large trees of almost any kind, in conifers such as spruce and Scots fir, and in tree ivy especially. It will also nest in tall hedges of hawthorn and blackthorn.

WREN

The wren builds in thick wall creepers, in tree cavities, on the trunks of large trees where there is a bristly growth of branches, in cavities in masonry and under the eaves of thatched roofs. Wrens seem to like material such as sacking as a nesting site. They will also use an open nesting box fixed in a creeper on a wall.

YELLOWHAMMER

The yellowhammer almost always nests in hedgerows and chooses a hawthorn, blackberry or blackthorn bush. In a garden it might nest in a hazel, myrobalan plum or beech hedge.

CHAPTER FIVE

Food, Water, Perches and Roosting

FOOD AND FEEDING TECHNIQUES

So much has been written in detail about feeding birds on bird tables that it is not necessary to go into the subject except in broad outline.

Perhaps the main problem with bird feeding is to prevent all the food being eaten by the birds that we wish to discourage, namely sparrows and starlings which will clear a bird table and eat the special and sometimes quite expensive food in a very short time. It is infuriating to see sparrows clinging to the plastic nut containers and extracting the nuts, while the tits sit hungrily by.

To attract the more interesting wild birds it is absolutely essential that the food supply should be constant. If the sparrows eat all the food within a short time of it being put out, it is unlikely that the more interesting birds will come to the feeding place, because the latter visit the garden only once or twice a day. If they find food always available their visits will become much more frequent, but if they find nothing they will go elsewhere. It is quite surprising how the blue tits, great tits, coal tits and greenfinches appear when they find a constant supply of nuts.

Sparrows on the other hand stay in a very restricted area, keep a sharp eye on any place where food is put out regularly and are on the spot immediately. They work in conjunction with starlings, virtually monopolizing most bird tables.

To keep the sparrows and starlings away is quite a problem, but it can be overcome to a certain extent by taking advantage of the fact that sparrows and, to a lesser degree, starlings, are by nature extremely wary and cautious. Although they can become very tame at times, they can quickly become equally wild if properly scared once or twice. The feeding platform and food containers should be placed within, say, half a metre of the sitting-room window. Sparrows will soon descend to the platform and start to clear up the food. When the platform is well monopolized by them, a newspaper should be suddenly banged on the window pane by a person concealed behind the curtain. If this is repeated several times sparrows will become very wary of the feeding table and will depart to some place where they can eat their food undisturbed, whereas birds such as greenfinches, tits, robins, blackbirds and thrushes will soon grow accustomed to being watched through a window at very close quarters and come readily to the table provided, of course, that the window is closed.

Most bird tables are put too far away from the house, on the false supposition that birds will be too frightened to come near the house. The great advantage of putting the table near the window of the sitting-room is that a really first-class view can be obtained of the birds visiting the table, and the table can be replenished from time to time merely by opening the window.

The food for a bird table should consist mainly of bread scraps, pastry, biscuits, cooked potatoes, cooked bacon rinds, bits of fat or meat, bones, nuts, acorns, raisins, apples, sunflower seeds, canary seed, pudding, potato crisps, stale breakfast cereals, old biscuits, bran, chicken and pig meal. Mealworms are the greatest delicacy, although these are expensive unless home-bred. However, robins, for example, will become finger tame with the aid of them.

To breed mealworms, find an old fish kettle or large biscuit tin and puncture the lid with holes to provide ventilation. Then place a piece of sacking on the bottom of the tin and cover it with a thick layer of bran. Add cut-up raw potatoes to provide moisture, and some slices of bread (in which the eggs are laid).

Food, Water, Perches and Roosting

Cover this with a layer of sacking, adding more bran, bread and potatoes. Repeat this process until the tin is nearly full. Then put in a few hundred mealworms, keeping the tin always at room temperature.

After a few weeks the mealworms will turn into pupae and eventually into black beetles of the type which used to be seen all too often in old-fashioned bakeries. The beetles will lay eggs in the bread and their eggs will hatch into mealworms. Food supplies should be topped up regularly.

It needs a certain amount of enthusiasm to breed mealworms successfully, but they are such a popular bird food that it is worth a try, especially if a warm, dry cellar is available.

One of the problems of feeding soft bills is that the bigger birds such as blackbirds are apt to fly off with a large piece of food and try to eat it elsewhere. Often the food is lost in undergrowth or taken away by members of the crow tribe. When snow is on the ground, it is invariably lost. To prevent this happening, it is best to devise some sort of feeder. One type has a wooden lid which fits on to a prism-shaped frame, covered by medium-sized mesh (about 3 cm.) wire netting. It is attached to the feeding table by wire struts at each corner and can easily be removed for cleaning purposes. Such a feeder can obviously only be used for bulky house scraps.

Food containers for tits, greenfinches, nuthatches and woodpeckers should hang free from a horizontal bar near a window and, like all food containers, should be suspended sufficiently high to be out of reach of ground predators.

Shredded coconut should not be given, as it swells inside the bird. Ordinary coconuts will often be left untouched by tits if there is an alternative supply of monkey nuts, brazil nuts or peanuts. The plastic containers of nuts are very satisfactory and there is very little waste from these.

Collared doves have now added themselves to our winter residents. They are mainly seed-eating birds but they will eat bread. In the summer they feed on weed seeds and live in gardens, usually on the outskirts of a town. In the winter they get together in flocks and find a constant source of food supply such as a corn mill. Wheat is their favourite food and it is a good plan to buy a sack of waste wheat to feed these attractive birds, although it is to be hoped that they will not multiply to such an extent that they become a nuisance.

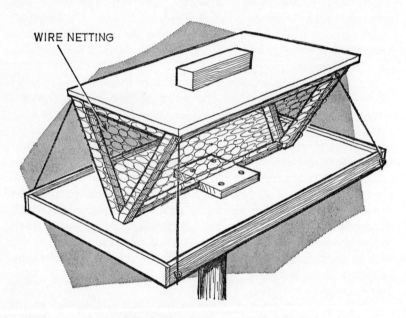

Wire mesh feeder for the bird table.

Great spotted woodpeckers are now frequent visitors to bird tables over most of Britain, and though they rob smaller birds of their food it is well worth trying to get these attractive birds to visit the feeding station. Perhaps the best way to do this is to put a squirrel-proof feeding tray, containing hazelnuts, almonds, walnuts, pieces of suet and a mutton bone, some distance away from the house near a tree which is preferably dead or partly dead—that is, one which a great spotted woodpecker would normally visit if it came into the garden. As soon as the new source of food supply has been found, the table can be moved gradually

Food, Water, Perches and Roosting

nearer the house so that these delightful, somewhat shy birds can be seen at close quarters. They will become quite tame but, as with most garden birds, this requires patience, time and quiet surroundings.

It is doubtful if green woodpeckers and lesser spotted woodpeckers ever visit feeding stations.

Great tits, in addition to their normal diet of insects, various nuts and birch seed, are particularly fond of beech mast, for which they have a vital need. (C. M. Perrin)

Some birds such as blackbirds, thrushes and mistle thrushes are not agile enough to cling to food containers and prefer their food on a table. This should be a simple wooden tray, preferably with a feeder, fixed on top of a wooden or metal post in such a way that mice, rats, cats and squirrels cannot climb on to it. The table should slope very slightly to allow water to drain off through a gap in the lowest corner. A roof over the bird table is quite unnecessary because birds always like to have an unobstructed field of vision when eating. Many bird tables are made with a roof over them, presumably to keep the food dry, but even so it usually gets wet because the rain drives in at the sides.

Birds such as hedge sparrows, chaffinches and wagtails seem to prefer to take their food on the ground. A hedge sparrow will usually hop round underneath the bird table eating the minute particles of food which fall on to the ground. For these dainty feeders it is advisable to scatter finely-powdered biscuits on the ground. Sparrows will not bother too much about these small particles if they can get hold of something bigger.

At a feeding table we can observe more easily the individual behaviour of various types of birds and can see how some birds, such as tits, have a greater capacity for learning than others.

Most birds have an inborn instinct to peck at things. Young chicks will peck as soon as they are hatched, and all birds learn things to their advantage by a trial and error system of pecking at objects. If they get a favourable reaction, such as the cream off the top of the milk bottle, they will remember to repeat the process and others of the same species will imitate them—for birds are good imitators. But if they get an unfavourable reaction, they will remember not to repeat the process.

There are a number of string-pulling experiments which demonstrate how tits can obtain food by a process of learning or by insight.

One such experiment is to tie a few nut kernels on to a light piece of string and suspend it so that it hangs about 30 cm. below the perch. The purpose is to see if the tit can learn to pull up the string with its beak, at the same time holding down the loops with its feet, until it reaches the nuts.

In another pulling experiment, a vertical nut container has several pegs in a row down one side. If the tit pulls out the correct peg, the nut falls into a tray below, like a fruit machine.

FEEDING IN THICK SNOW

Birds can stand a very low temperature provided they can get enough food. Their metabolism is rapid and small birds have to eat about a third of their weight per day to maintain it at a correct level.

Most birds build up a certain fat reserve during the autumn months and this reserve is sufficient to tide them over short periods of very severe weather; but if the weather is very cold with heavy snow, and lasts over a week, many birds of the thrush family, particularly redwings, may die of starvation. The stronger birds fly around in search of berries and will soon come to gardens. They are all very hungry and some of them will be tame enough to come to the table, while others such as fieldfares, redwings and mistle thrushes will hang about looking rather miserable in the snow at the end of the garden.

This is the time to make a second feeding place by clearing away an area of snow, either on the lawn or in an open place away from trees if possible. Clear out all apples from the store if they show any signs of decay and put them out for the birds. Boil up any doubtful-looking potatoes and throw out all scraps, including chicken carcases. In coastal areas seagulls may take a great deal of the food, but to some extent this can be avoided by dividing the food as finely as possible as seagulls like to pick up large pieces and make off with them. A herring gull can swallow a chicken drum-stick in mid-air without difficulty. It is a good

Food, Water, Perches and Roosting

idea to cover up such a feeding site at night with sacks; then, if it snows again, the sacks can be easily removed with the snow on them, thus keeping the feeding place clear.

Birds such as crows, magpies, rooks and jackdaws usually manage to find enough food even in the coldest weather, as they watch out for birds which are weak or starving and soon kill and eat them. The jackdaw and rook have adopted an interesting technique for getting a meal easily. They wait until a bird with a piece of food in its beak flies away from a bird table. They then dive-bomb the poor bird until it gives up its food and has to go back for more, when the process is repeated.

FEEDING IN FROSTY WEATHER

In very frosty weather insect-eating birds, particularly the blackbird, find a lot of food by turning over leaves, the insulating properties of which keep the underlying ground free from frost and allow insects and worms to be found. It is therefore a good idea, where space allows, to pile up some leaves in a sheltered part of the garden, where they will not be scattered by the wind.

Birds which are insectivorous, such as blackbirds, thrushes, robins, hedge sparrows, wrens and wagtails, have no difficulty in getting a living during the winter provided there is no severe frost or snow. Many of these birds stay out in the open countryside away from buildings if the weather is favourable. If a severe frost comes, they will move round the district looking for food. This is when they are in real need of a good meal and eventually come to the bird table. The jealous garden robin will even allow other robins to invade his territory. Wrens very rarely visit bird tables, but like to go through the garden sheds looking for spiders and other small insects. Seed-eating birds such as chaffinches, greenfinches, linnets, goldfinches and bullfinches are not so hard-pressed for food unless there is a heavy fall of snow which covers up all seeding heads of weeds. Hawfinches, which are only found in certain districts of England, manage quite well on the pips of wild berries.

Greenfinches will soon appear if a seed container is hung up, and they may be followed by chaffinches. Many linnets and goldfinches migrate for the winter and those that remain behind rarely visit bird tables. Bullfinches stay with us for the winter, going round from one place to another in search of buds. They will often start eating the buds of fruit trees in late December and can find a living on their own. They seldom visit bird tables, though this has been recorded in some areas.

The green woodpecker, which spends much of its time eating ants on the ground, suffers in severe frost or snow. Its numbers were greatly reduced by a severe winter in 1962.

NATURAL FOOD IN WINTER

As birds require a source of food which can act as an emergency supply when their normal food is not available, it is wise to plant shrubs and trees which are late fruiters rather than those which fruit in August and September when the food supply is plentiful. The berries of the hawthorn undoubtedly provide much winter food for blackbirds, thrushes, mistle thrushes, fieldfares, redwings and waxwings. It is obvious that if these trees are cut back every year there will be no berries, and unfortunately this is what is happening to our roadside hedges. Anyone who has a free-growing hawthorn in his garden will have a good supply of winter food for garden birds.

The berries of all types of holly tree are eaten during very cold spells by the thrush family, but they seem to be less popular than the berries of the hawthorn. Cotoneaster berries will remain on the branches until the severe weather sets in, and such berries are very attractive to waxwings. It is thrilling to watch these exotic-looking birds which come all the way from northern Europe to winter with us. Waxwings are fairly tame and will visit any garden in either town or country in search of berries. Pyracantha berries ripen in September, but as they are very popular with blackbirds there are often not many left when the severe weather appears. Crab apple trees are, of course, very useful, and the fallen apples will often last a long time on the ground and will be eaten by the thrush family and tits. The rowan tree or mountain ash is always a reliable fruiter, but its September berries have usually disappeared before Christmas. Beech mast is an important winter

Food, Water, Perches and Roosting

food of the great tit, chaffinch, brambling and wood pigeon. The seeding heads of rhododendron attract lesser redpolls and siskins. The latter are particularly fond of the seeds of the alder tree, which is usually seen growing beside a river. Most seeding weeds in severe weather will be eaten by goldfinches, bullfinches, greenfinches and linnets. Such weeds are dock, plantain, chickweed, thistle, shepherd's purse, teazle, groundsel and grass seeds.

ALL THE YEAR FOUND FEEDING

It is generally not necessary to feed birds all the year round, except in very built-up areas and in dry weather, when insectivorous birds cannot get enough food for their young. Those of us who have the time and interest, however, to put out a regular supply of mealworms, will be well rewarded because birds will become almost finger tame and completely trusting in the person who puts out their daily supplies, even though they will still retain their inborn fear of any other person.

WATER

Birds use water both for drinking and for bathing. Most garden birds like bathing—the starling in particular will take a bath however cold the weather is. Some birds, such as game birds, have a dust bath to clean their feathers. Pigeons and doves do neither. Birds never soak themselves completely, but dampen the ends of their feathers and then, using the oily substance from their preening gland, they clean each feather individually, particularly the flight feathers. They use the most ingenious antics to get the oily substance on to the parts inaccessible to the beak.

The ideal bathing and watering place for birds is a shallow cement pond holding not less than 60 litres of water, so that the risk of drying up is lessened. Shallowness at one end is highly important. Small birds like a depth of only about 1 cm. and larger birds a depth of 3 cm. so that the pond should have a gradual slope.

Once a bird has established a habit of drinking at one particular place, it will come regularly several times a day as long as the water supply is adequate. One method of keeping the water supply constant is to connect the down-pipe which drains the rain water off a shed or garage, to the pond. By this means, except in very dry spells, the pond will remain well topped up.

The siting of the pool is very important. It should be in fairly open ground in the sun, and not too close to any shrubs which may conceal a cat. The north end of the pool should have a small bank facing squarely to the south. In very cold weather, when the pond is frozen hard, the ice will thaw at the edge under this bank by midday, if there is any sun about, and will provide a valuable source of water for the birds. The siting of the shallow end does not matter provided it is not in a dip, which would prevent the birds from keeping a sharp look out for enemies while drinking and bathing. A well-appointed pool is a great asset to any garden and gives a centre of interest. Birds seem to be attracted to water in the same way as we are. Whether because of the aquatic flies and beetles which frequent small ponds or whether it is because they use it as a sort of meeting place, it is impossible to say.

A finely-set lawn spray will always attract the birds which like bathing. If used during long, hot, dry spells it will act as a substitute for rain and, in addition, it will bring worms to the surface.

Plastic pools are easy to put in, but are generally not very attractive to birds because they cannot get a proper grip with their claws, as they can on cement; but if small stones of about 6 cm. diameter, which don't slip about easily, are carefully laid on the bottom leaving a depth of a few centimetres, the birds will be able to obtain a better grip while bathing. Ornamental baths again are unsatisfactory, because they are usually small and dry up too easily, thus needing constant attention.

Birds require water during long, frosty spells when the ground is dry and hard, but when there is snow they quench their thirst with the small amounts of snow which stick to whatever food they eat. Chickens will always eat snow if they cannot find their water trough. A straw-filled sack placed over a corner of the pond on top of a board will help to stop the ice becoming very thick, or may even stop it freezing. A more certain method of stopping water freezing is to use a thermostatically-controlled immersion

Food, Water, Perches and Roosting

heater of the type used for aquaria. The container should be made of poor heat-conducting material, such as asbestos or plastic, and sunk into the earth to avoid heat loss. A pyramid-shaped rough stone should be put in the middle to prevent small birds from drowning themselves. Specially made drinking containers, with heating mechanism attached, can be purchased.

PERCHING POSTS

When a bird flies round the garden it usually goes to certain favourite perching places in various trees. These are used as singing places and also as vantage points to see if all is clear before flying elsewhere. A cock blackbird usually has three or four singing posts round the nest, using each in rotation to issue his warning to rival blackbirds to keep out of his territory. Robins, flycatchers, mistle thrushes and collared doves are all birds which have their favourite perching places in a garden. It is therefore a good idea to fix up an artificial perch in the shape of a T by using a rustic pole about 2 metres high with a short transverse pole on top. The best place to put it is near the drinking pool, where birds will use it as a preliminary perch to see that all is clear before they go down for a drink. If there are flycatchers about they will almost certainly use the perch, and will provide a good display of their aerial acrobatics from such a post. During the night, owls are very likely to use it as a perch from which to watch for mice. It is the simplest thing to construct and will make it possible to obtain a much better view of many garden birds.

ROOSTING PLACES

A sheltered and secure roosting place is essential for a bird's survival, and this applies more particularly during the winter months when a bird has to spend as much as 16 or 17 hours out of the 24 in its roost. During this time it is losing a great deal of body heat, especially in very cold weather. Birds will therefore choose a place where they are out of the cold wind and protected from rain. They sometimes need to alter their roosting site according to the weather or the direction of the wind, but usually as soon as the cold weather starts they select a communal roosting site permanently protected from north and east winds and with sufficiently thick evergreen foliage to keep off the rain and snow. They will use this roost throughout the winter, returning each day to their own special feeding grounds. As soon as nesting activities begin the cock bird roosts in a tree near his sitting mate.

Most garden birds roost at least 2 to 3 metres from the ground and near the top of the tree so that they can fly out quickly if disturbed. They avoid using tall trees, as these would catch too much wind. The ivy-covered trunks of trees are popular roosting places because here the bird can change its position according to the prevailing wind. Clumps of laurel, rhododendrons or thick

Perching posts are simple to construct and are a valuable asset for bird observation.

Food, Water, Perches and Roosting

evergreens about 3 metres high are also much used. Holly trees are quite popular, provided there are several trees together. Thick ivy growing on a south-facing wall is especially popular. Yew trees are less useful for roosting as the leaves are too small to keep off the rain. Certain types of cypress trees which have a very thick, vertical growth are almost weatherproof, and such trees are often used by hedge sparrows and robins for roosting. Fir trees, such as the spruce, do not give sufficient wind protection unless they are grown in clumps. An evergreen which has big, strong leaves, such as a rhododendron or laurel, particularly if grown against a sheltered wall, will always be popular as a roosting place. Bramble bushes provide very popular roosting cover except during the winter months.

A bird always tries to avoid getting wet at night because it has no chance of drying itself by flying about, and may lose so much body heat if it gets really soaked that it can die of cold. If it is raining or snowing when a bird goes to roost, it will get under a large leaf and thus keep dry; but if it is windy as well, the leaves will be disturbed and the bird will have an uncomfortably wet night. For this reason, birds always select a roosting place well sheltered from the wind.

It is still sometimes believed that birds sleep in their nests. With a few exceptions, this is quite untrue, although wrens will roost together in old nests in order to keep warm, particularly if the nest is in a very sheltered spot such as on a wall, and owls will sometimes sleep during the day in their nest holes to avoid being mobbed by blackbirds and other garden birds. Sparrows, however, roost regularly in their old nests and also in nest boxes. In one nest box scheme, a third of the boxes in a count of 200 were used by sparrows for roosting.

Young swallows and house martins, when they start learning to fly, will leave the nest for the whole day and will return to it again at night to roost. They will do this for perhaps ten days. It is therefore important not to pull down old nests too soon, particularly those of late-hatched nestlings which are often not ready to fly until mid-September when the weather can be quite cold and wet.

WREN ROOSTS

As already mentioned, wrens will roost in considerable numbers packed together for warmth in very cold weather. The ideal spot to put up a wren roosting box is in a porch adjoining a warm living-room facing south. An open nesting box filled with clean hay will be suitable if it is fixed, about 2 to 3 metres high, to an inner wall of the porch. It is quite amusing to count the number of wrens packing in on a late winter afternoon, and in very severe weather such a device may well save the lives of many of these charming little birds.

CHAPTER SIX

The Enemies of Birds

BIRDS HAVE more enemies than we perhaps realize, quite apart from the natural hazards of snow, severe frost, floods and, for some, the perils of migration.

The most dangerous time in a bird's life is when, as a fledgling, it leaves the nest. Its flight is weak and its perching ability not properly developed.

After leaving the nest, the young of most garden birds sit quietly for a few days, well concealed in a tree near the old nest, during which time their flight and tail feathers are growing rapidly. But if for some reason they are suddenly frightened, they scatter in all directions, some fluttering to the ground where they can fall an easy prey to a cat, rat or weasel. Others, being stronger, may perhaps manage to fly to another bush where they may survive, provided they keep off the ground. This scattering obviously causes much greater feeding problems for the parent birds and no doubt many young birds starve in this way.

Fledglings will also pop out of the nest a day or two before they would normally leave if they are disturbed by anything such as hedge-cutting or even by just being looked at. Once this has happened it is usually a waste of time trying to put them back in the nest, because they come out again. The best thing to do is to put them in an open-fronted box with wire netting (2·5 cm. mesh is usually suitable) across the front, putting the old nest inside.

The box must be fixed firmly in the nesting tree or nearby, and it is most important to see that it is out of the reach of cats who will pull the whole box down if they can climb up to it. The young birds will be fed through the wire by the parents and after about five days, provided they appear to be in good shape, they can be released.

If young birds can survive this dangerous time in their lives, they are then faced with further enemies which unfortunately are on the increase. It is therefore essential for anyone interested in bird survival to understand something about the ways of these enemies as well as the hazards which we ourselves create.

MAN

From very early times, birds of all kinds have been netted and snared by man as a form of food, but their numbers were not noticeably depleted because the existing human population was small and only destroyed such birds as it needed to eat.

With the advent of the shot-gun things became very different. In the latter years of the nineteenth century there was a great craze for shooting rare and beautiful birds for stuffing, and many of our most attractive birds such as woodpeckers, kingfishers, goldfinches, hawfinches, barn owls and many rare species were shot in order to satisfy this somewhat bizarre taste in ornaments. The craze for buying up these relics has come back, but very fortunately taxidermists are few and far between and it is now an expensive item to have a bird stuffed. Birds of prey such as eagles, ospreys, peregrines, hobbies, harriers, merlins and sparrow hawks have suffered great reduction of numbers as a direct result of the shot-gun. Many of these birds also died in gin-traps and pole-traps, but these very cruel forms of trapping are now illegal.

Today shot-guns are relatively cheap and there is much indiscriminate shooting by those who like to 'have a go' without knowing one bird from another, and who are prepared to shoot almost anything that runs or flies. In some Continental countries thrushes, blackbirds, larks, chaffinches, coal tits, sparrows, buntings and many other birds are looked upon as delicacies of the table and are shot or netted all the year round. Air-guns can play havoc with garden birds when in irresponsible hands.

Egg collectors are not quite so numerous now, but they are still sufficient in number and in cunning to mark down and rob the nests of the very rare birds which we particularly want to preserve.

The gradual urbanization of the countryside poses the greatest threat. Forests of cement are no substitute for trees and hedges, and all over the country our wild birds are being pushed out of their previous surroundings.

The great progress in all branches of science is a potential threat not only to the lives of the lower creatures but to human lives as well—which is perhaps a good thing for, if we know we are all in it together, more care may be taken in future over how we apply our ever-expanding scientific knowledge.

POLLUTION

This is a subject about which much has been written recently in relation to the sea, rivers and agricultural land. It is hardly necessary to say that we must also be very careful about what we use in our gardens.

Man has already started to exploit nature, and in many instances the results of his exploitation have been not only unexpected but somewhat alarming. Biological processes are exceedingly complex and any exploitation has to work in the total context of the environment, namely man, animals, birds, insects, air, water, soil and plant life. Even a minor disturbance may produce a host of changes, both large and small. We cannot possibly foretell accurately from a laboratory experiment all the effects a certain chemical will have when it is used outside the laboratory in the complex environment of nature. When an error is made it can often be corrected, but this may not be before it has done severe and sometimes irreparable damage to living creatures, plants, insects or soil. So far garden birds do not seem to have been affected, except that by using sprays we may be cutting down the supply of insect food which is so important to birds feeding their young. Possibly the excessive use of insect sprays (which is likely to be greatest in very small gardens in built-up areas) may have a restrictive effect on clutch sizes. Some of the chemicals formerly used in gardens, such as aldrin and dieldrin, are both persistent and extremely poisonous. Others, such as chlordane, endrin and parathion, are also poisonous. The use of BHC in the garden is likely to be dangerous during May, June and July when young birds are being fed.

The alarming findings of excessive quantities of DDT even in polar bears and also in humans who have eaten fish and game in Canada, and the possible cancer-forming properties of this chemical, illustrate how we should not apply our scientific knowledge. An even more recent and equally alarming discovery is that a defoliant—2,4,5-T—used for killing brambles and undergrowth, causes death by paralysis of any living thing which gets a big enough dose; and, even worse, it has foetal-deforming properties.

URBANIZATION

As mentioned earlier, the encroachment of bricks and mortar into the countryside cannot fail to present a very serious threat to the livelihood of all birds except the starling and sparrow. Those who plan new estates should not ban the growing of hedges round gardens or the erection of fences. It doesn't help the birds and it may be that it doesn't help us, either. Open-plan living is all right for some people but it can be carried too far, and psychologists say, not surprisingly, that we are far more prone to anxiety states and emotional disorders if we live too much on top of one another.

Those who plan factories and public parks should endeavour to plant trees which are useful to birds. The hawthorn, for example, is very cheap compared with other flowering shrubs, looks very colourful when grown as a standard tree in a public place and will certainly be used by garden birds for nesting. Its berries in the winter are an important source of food for resident birds and winter visitors. Clumps of laurel, holly and rhododendron should also be planted for protection against weather and for nesting and roosting.

Again, in most parts of the country road hedges are now trimmed by mechanical means. The tall, thick hedge in which a great number of birds regularly used to build their nests is now reduced to a meagre, thin little hedge in which no self-respecting bird will nest. Furthermore, hedge-trimming, weed-cutting and weed-spraying are now often carried out in the height of the breeding season, where formerly hedge-cutting and ditch-clearing were traditionally jobs to be done at the back end of the year. There is no doubt that the new mechanical method of hedge-cutting has caused a vast reduction in the numbers of birds nesting in hedgerows.

THE MOTOR-CAR

An ever-increasing menace to all birds is the motor-car. The number of birds killed on the roads each year runs into millions, according to a recent estimate. The greatest slaughter occurs in the early mornings of late spring and summer when birds fly on to the roads, probably to get grit and, not being fully alert after a night's roost, are killed by cars. It is frequently the young and inexperienced birds which succumb, for they have not had time to learn that a car is a highly dangerous machine. When birds are on the road, a quick toot on the horn will often save a life. Many people will not bother to do this and think it is of no matter to kill a few birds. They may even deliberately try to mow them down.

Quite apart from the death on roads caused by motor-cars, the invasion of the countryside during the breeding season, particularly at weekends, by thousands of cars containing picnickers and day-trippers inevitably causes great interference with the successful breeding of birds. Bird photography has in the past done some harm where it has been pursued too enthusiastically, causing rare birds to forsake their nests, but it has now been controlled for certain rare species. It has done a great deal of good by giving the public information about the more interesting and intimate side of bird life. Bird-watching with binoculars is harmless enough when done by experienced people, but when parties of amateurs take to the countryside a great deal of disturbance can be inadvertently caused to bird life. It is unfortunate that it is always the rarest and shyest species which attract the most attention, and this must inevitably disturb the very birds which should be left in peace.

Motor boats, speedboats and helicopters are likewise invading the territories of birds who like solitude; and now, owing to the combustion engine, there are very few places left in the British Isles which are inaccessible to man. Such interference with nesting activities is serious, and more and more nature reserves must be created if we are to maintain the numbers of these rarer birds.

FOUR-LEGGED ENEMIES

Cats

The domestic cat probably originated from ancient Egypt and came via Europe to Britain hundreds of years ago. It is not a direct relative of our native wild cat which is now found only in the remote parts of Scotland.

Our domestic cat seems to enjoy the best of two worlds. It gets all the advantages of civilization—food, warmth and general protection from enemies—together with unrestricted, easy hunting whenever it wishes to satisfy its urge to kill. For cats are natural hunters and enjoy it. Some are good hunters and others are bad. They kill whether hungry or not and if they are not hungry they will play with their prey for some time before killing it. Female cats are usually the most efficient hunters and will kill rats, mice, weasels, rabbits and birds. They never restrict their killing to mice alone, as some people imagine, and they are thus a great deterrent to bird life in a garden. A cat prowling about at night makes all the birds uneasy and causes them to move elsewhere; and though the cat may catch only a few birds, it will reduce the number of bird visitors to a garden. Domestic cats which have gone wild and have to live off the country are a serious menace.

In the nesting season, cats will cause great destruction of bird life by both frightening the sitting bird off her nest at night and by killing off the young birds before or after they leave the nest. Cats find a nest quite easily by watching the parent birds going to and fro and they can climb any tree or shrub provided the branches will support their weight. They do not usually climb higher than

The Enemies of Birds

3 metres when hunting, as from this height they can safely jump to the ground in a case of emergency—for it must be remembered that, when hunting, the domestic cat becomes almost a wild creature with all the instincts for self-preservation.

When a cat reaches a nest it will usually claw the nest over and tip out the contents. If it is full of young birds ready to fly this will add to the cat's entertainment.

Those who keep cats and let them out at night should remember that in built-up areas cats will wander from one garden to another, trespassing in the gardens of people who prefer birds to cats and generally creating a disturbance to bird life. A partial remedy for this is for cat lovers to let the cat out only during the day from mid-April till the end of July when the nesting season is at its height; if it is well fed, the cat will probably find a sunny corner and go to sleep. A cat bell on a collar is quite useful, not because it gives a warning to the birds, but because it puts the cat off its hunting. Just as the cat is making a silent spring on its prey, the bell gives an unexpected tinkle, which must be very annoying to an animal which relies so much on the element of surprise in its hunting tactics.

The best way to keep cats out of a garden is to use 5 cm. mesh wire netting. For a hedge about 2 metres high, it would be necessary to use netting only 1 metre wide, which should be placed along the bottom of the hedge and pushed well in so that the twigs grow through the wire and cover it up.

Cats can jump a 2-metre wall if they can get a run at it. A length of wire netting about 50 cm. wide put along the top of a wall or fence will stop them getting over. They can easily be prevented from climbing smaller trees by an encircling wire stop around the trunk. Cats climb by embracing the trunk closely with their front legs and pushing up with their hind legs. Thus they cannot let go with their front claws for more than a fraction of a second, and if they try to circumvent any obstacle in their upward path they will certainly fall.

Dogs

It is rare for dogs to become destroyers of nesting birds.

Labradors, with their keen scent and voracious appetite, may occasionally find and devour the eggs of a sitting pheasant or other ground-nesting bird but, generally speaking, dogs are not interested in hunting birds in the way cats hunt birds. Garden birds realize this and are therefore not unduly worried by the presence of dogs from a nesting point of view, knowing that there is less likelihood of cats being around if there is a lively, yapping dog in the garden. A dog of the terrier breed will tend to keep away ground predators such as foxes, weasels, stoats, rats and mice.

Dogs will usually avoid areas sprinkled with a fox repellent, and it is therefore wise to use this mixture where there are nests on the ground.

Unfortunately, in a garden where there are dogs (and, likewise, small children), birds are unlikely to become tame because they are disturbed if they see anything rushing around. Puppies, for instance, will often make a sally at a blackbird or collared dove feeding on the lawn, just for the fun of it. Sudden, unexpected movements make birds more wary.

Foxes

In spite of the scarcity of rabbits since the introduction of myxomatosis, the fox has probably not decreased in numbers to any great extent. There are in fact many more foxes about than we realize, but because they still retain their nocturnal habits we rarely see them unless we get up very early in the morning and keep a special watch in a likely area. In many places, particularly in residential districts where they are not persecuted, they have greatly increased in number. In the suburbs of London, for example, foxes are quite numerous and breed unmolested, living on mice, rats, chickens, birds, pet rabbits and garbage on the rubbish tips. They will eat almost anything, including a cat if really hungry. Living in the suburbs (such as Hampstead, Wimbledon or Beckenham) with no pest control officers or packs of hounds to chase them, they have an easy life.

In those hilly parts of the country where pine forests have been planted, new strongholds for the fox have been created. These hill foxes used to be controlled by hunting or by the use of working terriers, but they are now unapproachable in these almost impenetrable pine forests. A big surplus of foxes in the remote districts where our few remaining birds of prey can still be found may well create a food shortage and cause the birds of prey to move elsewheres.

There is no doubt that in the nesting season eggs and young birds form quite a considerable part of the fox's diet. When there are hungry cubs to feed and no rabbits about, a fox will take a bird incubating her eggs from any nest within his reach, plus all the eggs or young. Being an agile climber he can get up to quite considerable heights, and may cause a lot of incubating hens to forsake their nests by disturbing them at night, even if he cannot reach the nests.

Unfortunately there is very little one can do to keep foxes away. A fox can get over, under or through almost any barrier except a fox-proof fence, which is an expensive item. An oil lamp hanging from a tree at the end of the garden will probably keep a fox away quite effectively for a month. The position of the lamp should be changed occasionally. Fox repellents, sprinkled round a nest, are useful. Anything which suggests a trap makes a fox very suspicious. Pieces of metal or tins hammered to a stake and placed near a likely entrance through a hedge may make a fox turn back. The metal should be handled periodically to keep the human scent on it. The fox is a good swimmer and water will not stop him from taking ducks nesting on islands in ornamental pools. He is also a master hypnotist, and can often make his prey come to him if he cannot catch it by his usual methods. Members of the duck and crow tribe are particularly fascinated by foxes, and if they see one behaving in a peculiar manner or in any trouble they will quickly be on the scene to inspect.

The curiosity shown by wild duck about foxes has led many to their destruction in the following way. A mongrel dog, which has the same shape and colour as that of a fox, is trained to lure wild duck into a decoy or 'pipe' which consists of a small stream covered over by wire netting arched over the top. The stream leads off the main feeding pond. The foxy-looking dog appears at the mouth of the decoy and then disappears into the decoy to re-appear further up, behind specially-constructed hurdles. The duck, with their insatiable curiosity, swim into the trap and keep following as the dog repeats its act. Eventually the dog-handler appears behind them at the entrance of the 'pipe' and drives the duck towards the narrow, upper end of it where they are caught in a net. This method was first used in the seventeenth century for catching duck for the table, but is now used only for ringing purposes.

Under normal conditions the fox would present no menace to the lives of smaller birds, but since man has again interfered with the balance of nature by introducing myxomatosis the fox has had to change his diet to survive. There is no doubt that birds now make up a large part of his diet and will continue to do so.

Grey Squirrels

The charming ways of these animals are in sharp contrast to their methods of getting a living. They hunt both on the ground and in the trees, and whenever they find a nest they will devour the contents, eggs or young. They also eat fruit, nuts and acorns and will readily rob a bird table or nest box if they can get into it.

It is a long time since grey squirrels were first introduced into Britain; but they must nevertheless still be regarded as a foreign species, and be discouraged in a most positive way if we wish to preserve our bird life.

Weasels, Stoats and Mink, Badgers

The weasel is a charming but ferocious little acrobat who will rob any nest he comes across either on the ground or in a tree.

The stoat, though less ferocious, has similar hunting habits. Both these creatures kill vast numbers of rats and mice which otherwise might become plagues. Weasels and stoats therefore play a very important role in the balance of nature and should be left unmolested by man. They usually breed twice a year and it is unlikely that their numbers would ever increase sufficiently to allow them to become a nuisance.

Escaped mink are now to be found in certain parts of the country. These creatures are fierce, uninhibited killers, destroying anything that lives in a tree or on the ground. A district unfor-

The Enemies of Birds

tunate enough to be occupied by escaped mink will soon have very little wildlife left, for they kill for the fun of it as much as for food. Anyone who sees a mink should report it to the local pest controller.

The badger, a delightful animal, is not likely to do any harm worth mentioning to garden birds, and should be encouraged.

Rats and Field Mice

Both these animals are excellent climbers. The field mouse, a much bigger and more handsome fellow than the house mouse, is a robber of small birds' nests. Rats do not do a great deal of damage to tree-nesting birds because during the nesting season there is usually plenty of food on the ground and climbing exposes them to a certain risk. They would always take the contents of a nest built on the ground.

Field mice, on the other hand, spend a large part of their time climbing in hedges, bushes and even big trees. They often eat the eggs of birds and may sometimes convert the nest into one for themselves.

Hedgehogs

These delightful animals are virtually harmless to bird life. If, in their quest for insects, they came across a willow wren's nest or that of some other ground-nesting bird, they would certainly devour the contents, but such occurrences are few and hedgehogs are mainly insectivorous in their diet.

FEATHERED ENEMIES

Crows

The crow tribe in Great Britain consists of ravens, carrion crows, hooded crows, rooks, jackdaws, magpies, jays and the rare chough.

The raven and chough, both of which live in remote coastal areas and have in any case different feeding habits from the rest of the crow tribe, need not be considered here.

Carrion crows and their close relations the hooded or grey crows will eat and devour any eggs or young they can find and are perhaps a greater menace to ground-nesting birds such as larks, meadow pipits, game birds, waders, ducks and gulls than to smaller tree-nesting birds. The grey crow is distributed throughout northern Scotland and Ireland. The carrion crow is rarer in Scotland but fairly common in England. Carrion crows easily become urbanized, but they tend to keep rather aloof from gardens and will live and build in parks and waste places.

Rooks and jackdaws are mainly interested in insect food and usually feed together in flocks. Both species dislike crows and magpies. Jackdaws will gang together during the nesting season to drive these robbers away from their nesting area, for they know that the young jackdaws will be subject to attack when they emerge from their nest hole. If rooks or jackdaws find a nest on the ground they will certainly eat the eggs or young birds, but they are not professional nest-hunters like the magpie and jay.

Magpies

Of all the crow tribe, the magpie is the most cunning and most destructive of bird life. It will take the eggs and young of any bird whose nest it can reach, and the only birds immune to its attacks are those that build in holes and crevices, such as sparrows, starlings and tits.

Since the last war it seems to have become much more common in residential areas, to which it has adapted itself very quickly. The magpie population used to be mainly confined to country districts, where numbers were kept under control by gamekeepers and farmers who knew all about its bad habits. There are now very few gamekeepers, and farmers no longer need to bother themselves about magpies taking their young chickens and ducks because these are now hatched and reared under cover, whereas they were formerly hatched under a hen and put in a coop out in the open where they could fall an easy prey to magpies.

The result is that in areas where magpies were never seen, they are now becoming relatively common. They have invaded urban and semi-rural districts where they find life extremely easy with ample food supplies throughout the year, and being most efficient

The Enemies of Birds

breeders, their numbers are increasing. They are such expert thieves that people rarely see them at work and find it hard to believe that these beautiful birds are so destructive to bird life.

The magpie works with a fixed tactical plan. While the hen magpie is incubating her eggs the cock goes off hunting for eggs and young birds. As a good burglar he 'cases the joint' before action. This is done by sitting on certain high trees and watching the comings and goings of the nesting birds in gardens and hedgerows. Having marked down certain nests for destruction, he comes in very early in the morning when he can work undisturbed and quite fearlessly robs the nests he has marked, wherever they are situated. He will take the young out of a blackbird's nest right under a dining-room window and the eggs of a flycatcher in a creeper on the garage. He will take the eggs from a robin's nest in the ivy behind the shed and the young of a chaffinch in the hawthorn at the end of the garden. All this is done very quickly, without fuss or noise. Even the blackbirds do not give an alarm call, as they will for a jay or an owl. No person would be aware that a magpie has been about, unless he happened to be watching the progress of the nests in his garden.

The cock magpie will not come back again for some days. He has got plenty of other nests marked down for destruction. After about a week the birds whose nests have been robbed will have built again and there may be two or three eggs in the nests. These will again be taken. This goes on throughout the breeding season, the garden birds providing food for the rapacious magpie.

During one May I kept a close watch on 43 nests—those of blackbirds, thrushes, hedge sparrows, robins and one whitethroat's. Of these only one blackbird brought up its young safely; the remaining nests were systematically robbed of their contents, either eggs or young, almost certainly by magpies. There were few cats in the neighbourhood, and it is always possible to trace the work of cats because the nest is invariably tipped over, whereas a magpie never disturbs the nest. A fox will also tip the nest over. The garden in question is thickly planted with evergreen trees and is on the edge of open country infested by magpies. It was therefore particularly subject to their raids.

If we wish to maintain the population of our smaller birds we must keep down the population of magpies. The most humane way to do this is to destroy their nests, which are fortunately very large and easily visible because they are built early in April when the trees are bare. Destroying a magpie's nest is not at all easy, as it is a very solid structure and usually fairly high up in an inaccessible tree or among thick thorns. If it is accessible with a ladder, the branch supporting the nest should be sawn through. This will, at least, scare the magpies off for the season, though they will go elsewhere to build again. The advantage of sawing through the branch rather than poking the nest down is that the magpies cannot build again, as they sometimes do, in the same nesting site. The nest should be destroyed at the end of April, when there is no possibility of any young birds having hatched.

In an attempt to keep them away from a nest, it sometimes helps if a piece of flapping black and white material, shaped like a magpie, is hung up about 2 metres away from the nest. This may only be effective for a week or so, but it is worth trying.

Jays

The jay is a much shyer bird than the magpie and does not seem to have established itself in built-up areas to any extent. It is very much a bird of the woodlands, where it certainly destroys the eggs and young of birds, but it rarely visits the average-sized garden on hunting expeditions and need not, therefore, be considered here as an enemy of any real danger to garden birds.

House Sparrows

The cheerful, cheeky sparrow must unfortunately be numbered as one of the enemies of other garden birds because it is not only greedy and aggressive but also such an efficient, prolific breeder that it tends to keep other birds away by sheer weight of numbers. House sparrows breed so successfully because they choose inaccessible nesting sites in buildings and raise as many as three broods a year.

Where food is concerned, they can live on an insectivorous diet or a seed diet. This gives them a great advantage over other garden birds, which are much more selective in their diet.

The Enemies of Birds

In addition, a sparrow, though only a small bird, is extremely strong physically, with a powerful beak and very well-developed flying muscles. It uses this strength to get what it wants.

Not only do sparrows monopolize the food supply, but also the nesting sites. They will evict a house martin from its carefully-built mud nest or harry a blue tit trying to build in a tit-box. I have seen a sparrow pulling out a flycatcher's nest, and another building on top of a swallow's nest, having first thrown out the eggs.

Sparrows have become parasitic on man and they are rarely seen far from buildings except during the harvest, when they settle in their hundreds on ripening corn. They are quick to learn, wary and suspicious, with a highly-developed sense of self-preservation. They are also to a certain extent parasitic on other birds. If, for example, they see a robin on the lawn with a tit-bit, several sparrows will arrive as if from nowhere, and if the tit-bit is to their liking they will take it; similarly, if a thrush is cracking a snail on a stone, a sparrow will invariably be around to see what sort of food the thrush has found.

Anyone therefore who wishes to have interesting birds in his garden should keep the sparrow population down as far as possible. The best way to do this is to block up all nesting holes in the house, wire over all drainpipes and destroy any nests. This is well worth doing, because one of the chief causes of blocked and broken drainpipes and damp walls is the sparrow's nest, which is often built in the swan neck of a roof drainpipe. When it rains heavily the nest is washed down the pipe. Another nest is then built and this suffers the same fate until there is a complete blockage of the down-pipe. If there is a severe frost the pipe splits and a great deal of expense can be involved in repairing or replacing it. Sparrows will also nibble away any loose mortar under a roof to make a nesting hole, and will pick the cement away from under the roof tiling at gable ends, thereby letting in the rain. It is unnecessary to say what they will do to a thatched roof.

They are related to weaver birds and can, if necessary, build a domed nest in a tree. It can usually be recognized as a rather ragged yet highly efficient construction of dried grass, old bits of string, feathers, newspaper and any kinds of odd materials.

Sparrows are estimated to be now the most numerous bird in the world. They have spread north to the Arctic circle, and it is fascinating to think how they have adapted themselves to these extreme temperatures. In certain intelligence tests they have been rated equal to the white rat. There is certainly no danger of a sparrow shortage, however much we restrict their breeding activities.

Starlings

This bird, like the sparrow, owes its rapid expansion in many countries of the world to the fact that it always builds its nest in inaccessible places and can eat almost any type of food. Starlings always seem to be hungry; when food is put out on a bird table, not one or two starlings appear, but perhaps a dozen, and the food quickly disappears. Their voracious appetites and expanding numbers are likely to prove a menace to garden birds, especially in a dry summer when insectivorous food for young nestlings is scarce.

Although they have many pleasant characteristics, these birds should be discouraged. We have encouraged them quite long enough—for whenever we erect a new building we are, in fact, putting up an artificial nesting device for them, as well as for sparrows, usually at the expense of the more interesting garden birds.

Starlings are cavity nesters and will build in any hole or chimney, but they never build a nest in the branches of a tree. They can cause considerable damage to houses by blocking up gutters and drains with their nesting material, and it is therefore a wise thing to stop up all holes and put wire over chimney tops.

Hawks

The kestrel is the hawk which is most often seen hovering over an open piece of ground, watching for the mice and beetles which form the main part of its diet. It will occasionally kill birds if there is a shortage of its normal food; in London kestrels feed largely on sparrows.

The sparrow hawk is still quite rare, largely due to its preying on

The Enemies of Birds

animals or birds which have eaten seed contaminated by pesticides and also because of persistent persecution by gamekeepers. Fortunately, its numbers are now increasing again. As its name suggests, the sparrow hawk does prefer to eat sparrows, and if there were more sparrow hawks perhaps we should not be so overwhelmed with sparrows. It is a noble bird and one which we should do everything to encourage.

Owls

Both the little owl and, more particularly, the barn owl appear to be much less common than they were. This again may well be due to poisonous seed dressing eaten by the rodents upon which owls prey. The tawny owl for some reason does not seem to have suffered as much as other birds of prey. Barn owls feed almost exclusively on small rats and mice. Tawny owls and little owls will take young birds as well as small rats, mice and beetles.

The harm done to birds during the nesting season by owls is infinitesimal compared with the damage caused by magpies, cats and squirrels, and owls should be actively encouraged as they kill off a great number of rodents.

Hawks and owls are included in this section because people may mistakenly regard them as a possible menace to garden birds. There is little danger of this, their present numbers being what they are.

CONCLUSION

Bird Conservation

FOR MANY of our rarer birds, the time has long since passed when they could survive on their own, without some form of special protection. The time will soon come when the more common birds of the hedgerow and garden will need protection, too, as the relentless destruction of their normal habitat continues.

This may sound like an alarmist call, but it is not; for if we continue at the current rate to encroach on the natural breeding grounds of birds, many species will suffer a severe decrease in their numbers. In brief, our many attractive garden birds are retiring slowly but surely before an army of sparrows and starlings who come like camp followers with the advance of man into the countryside. These camp followers are gregarious, and don't mind overcrowding. They are quick adaptors and sufficiently aggressive to take over new territory without any difficulty at the expense of the birds who lived there before.

The British Isles are still rich in bird life and have a climate which always attracts a great number and variety of birds. Being on one of the great migration routes from the north to the south, we are lucky to have many interesting winter visitors from the north. In the spring, we have many migrants from the south which come here because the climate is so suitable for breeding. Our rainfall is sufficient to guarantee the hatching of many flies and insects and to ensure a strong growth of weeds, shrubs and trees during the breeding season.

Throughout the year we have our own native birds which stay with us, fair weather or foul: the robin, wren, thrush, mistle thrush and blackbird—all great songsters—the unassuming little hedge sparrow, the tits, woodpeckers and the gaily-coloured finch family. These and many others make up our heritage of native birds. We cannot expect them always to be there unless we make provision for them. If we can only realize that, unwittingly, man above all else is the greatest enemy of all his fellow creatures, then we are some way to removing this very real threat to their existence. It is equally important to realize that man, if he understands the needs and habits of his fellow creatures, can, if he so chooses, be their greatest ally.

Fortunately, things are moving in the right direction. The teaching in schools on natural history has been extremely sensible and forward-looking and the younger generation take a great interest in the subject. Excellent television films have unquestionably been a major factor in showing us the charm of birds and animals and has stimulated a healthy interest in them. There is now much more emphasis on preservation, and though shooting will always remain an outlet to satisfy the natural aggressiveness and pent-up frustrations of man, many restrictions and penalties are being imposed to prevent the extinction of rarer species.

Birds have a very strong instinct for survival and are quick to take full advantage of any favourable conditions that we create for them. The balance between survival and death is often quite small, and the factor for tipping the balance one way or the other is invariably due to the action of man. Bird populations are certainly affected adversely by prolonged and severe cold weather and by storms and other natural hazards during migration, but they can recover from such set-backs; given a few good breeding seasons the survivors, being fitter and stronger than those who succumbed, soon bring the numbers back to normal.

But the constant interference by man, as opposed to the intermittent interference by nature, is something with which birds cannot cope and it is bound to cause a decline in all bird populations except those of sparrows and starlings.

Bird Conservation

There is little we can do for the shyer species such as larks, pipits, wheatears and many others who live away from human habitation, except to create sanctuaries. Much has already been done in this direction but we still need many more with greater restrictions on entry by the public. The high price of land means that sanctuaries are often made where the soil is infertile, rocky and treeless—territory which can only be inhabited by ground-nesting birds. Our more familiar songbirds need a much more fertile environment with trees, hedges and cultivated land in order to breed successfully.

Those of us who are interested in bird survival can each make a definite contribution even if we have only a small garden, for there is nothing more certain than that, given the right conditions, garden birds prefer gardens and will thrive in them—giving us, at the same time, much pleasure.

A little private enterprise or 'do it yourself', as opposed to leaving the problem to the powers-that-be, is essential if we wish to prevent what may be a common species today from becoming a rare one in the near future. We can most certainly help these birds and try to tip the balance in their favour not only by feeding them in bad weather but also by providing them with nesting sites, artificial or natural, and by seeing that they are left unmolested. The satisfaction of seeing a young brood of birds emerge safely from a purpose-built nesting site can be sufficient reward for all the work which has gone before. Difficulties and disappointments will arise despite every precaution, but this is the way of nature—and success is not worth having unless it is difficult to achieve.

NOTES

NOTES

NOTES